THE HIGHWAY CODE

AA Publishing

Contents

Produced by AA Publishing.
© Automobile Association Developments Limited 2004

Crown copyright material reproduced under licence from the Controller of HMSO and the Driving Standards Agency.

ISBN 0 7495 4436 8

Published by AA Publishing (a trading name of Automobile Association Developments Limited, whose registered office is Millstream, Maidenhead Road, Windsor, SL4 5GD; registered number 1878835).

A02329

The AA's web site address is *www.theAA.com/bookshop*

The contents of this book are believed correct at the time of printing. Nevertheless, the publishers cannot be held responsible for any errors or omissions or for changes in the details given in this book or for the consequences of any reliance on the information provided by the same. This does not affect your statutory rights.

Colour separation by Keene Group, Andover
Printed by St Ives, Andover

For more information about The Highway Code visit
www.thehighwaycode.gov.uk

Studying *The Highway Code* thoroughly is just one of the things you must do in order to help you pass your driving test. Before you apply for a full driving licence you must get to know *The Highway Code* and pass two driving tests – the Theory Test, including Hazard Perception, and the Practical Test. The Theory Test was introduced in 1996 to check that drivers know more than just how to operate a car, and the Hazard Perception element was introduced in 2001 to test learner drivers on their hazard awareness skills. This section explains the format of both parts of the driving test. It is followed by the complete *Highway Code* on pages 11-105.

You are strongly recommended to prepare for the Theory Test and Hazard Perception at the same time as you develop your skills behind the wheel for the Practical Test. Obviously, there are many similarities between the two tests and you need the experience of meeting real hazards in order to pass the Hazard Perception element of the Theory Test.

It is all about making you a safer driver on today's busy roads. By preparing for both tests at the same time, you will reinforce your knowledge and understanding of all aspects of driving and you will improve your chances of passing both tests first time.

The Practical Test was extended in September 2003 to include Safety Vehicle Check questions that test the driver's ability to carry out basic procedures to ensure that the vehicle is safe to drive. Checks include the tyre tread and oil level.

WHAT TO EXPECT IN THE THEORY TEST

You will have 40 minutes to complete the questions in the test, using a touch-screen to select your answers. The test is a set of 35 questions drawn from a bank of more than 1,000, all of which have multiple-choice answers. In order to pass the test you must answer a minimum of 30 questions correctly within the given time. The Government may change the pass mark from time to time. Your driving school or the DSA will be able to tell you if there has been a change. The questions appear on the screen one at a time and you can return to any of the questions within the 40 minutes to re-check or alter your answers. The system will prompt you to return to any questions you have not answered fully.

Preparing for the Theory Test

Read and get to know *The Highway Code*. You'll find that many of the theory questions relate directly to it. Other useful reference books for this part of the test include the AA's *Theory Test* which contains all of the official questions and answers from the Driving Standards Agency.

How to answer the questions

Each question has four, five or six possible answers. You must mark the boxes with the correct answer(s). Each question tells you how many answers to mark.

Study each question carefully, making sure you understand what it is asking you. Look carefully at any diagram, drawing or photograph. Before you look at the answer(s) given, decide what you think the correct answer(s) might be. You can then select the answer(s) that matches the one you had decided on. If you follow this system, you will avoid being confused by answers that appear to be similar.

WHAT TO EXPECT IN THE HAZARD PERCEPTION TEST

After a break of up to three minutes you will begin the Hazard Perception part of the test. The Hazard Perception test lasts for about 20 minutes. Before you start you will be given some instructions explaining how the test works; you'll also get a chance to practise with the computer and mouse before you start.

Real road scenes feature in the video clips in the Hazard Perception test

Next you will see 14 film or video clips of real street scenes with traffic such as cars, pedestrians, cyclists etc. The scenes are shot from the point of view of a driver in a car. You have to notice potential hazards that are developing on the road ahead – that is, problems that could lead to an accident. As soon as you notice a hazard developing,

click the mouse. You will have plenty of time to see the hazard – but the sooner you notice it, the more marks you score.

Click the mouse when you spot potential hazards – the pedestrian crossing the side road and the cyclist approaching a parked vehicle (ringed in yellow). Click again as the hazard develops when the cyclist (ringed in red) moves out to overtake the parked vehicle

Each clip has at least one hazard in it – some clips may have more than one hazard. You have to score a minimum of 44 out of 75 to pass, but the pass mark may change so check with your instructor or the DSA before sitting your test. (Note that the computer has checks built in to show

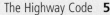

anyone trying to cheat – for example someone who keeps clicking the mouse all the time.) Be aware that, unlike the Theory Test questions, you will not have an opportunity to go back to an earlier clip and change your response, so you need to concentrate throughout the test.

Preparing for the Hazard Perception test

Who do you think have the most accidents – new or experienced drivers? New drivers have just had lessons, so they should remember how to drive safely, but in fact new drivers have the most accidents.

Learner drivers need training in how to spot hazards because they are often so busy thinking about the car's controls that they forget to watch the road and traffic – and losing concentration for even a second could prove fatal to you or another road user.

> **You have to pass both the Hazard Perception test and the Theory Test questions. At the end of the test you will be told your scores for both parts. Even if you only failed on one part of the Theory Test, you still have to take both parts again next time.**

Proper training can help you to recognise more of the hazards that you will meet when driving and to spot those hazards earlier. So you are less likely to have an accident.

Your driving instructor has been trained to help you learn hazard perception skills and can give you plenty of practice in what to look out for when driving, how to anticipate hazards, and what action to take to deal with hazards of all kinds.

You won't be able to practise with the real video clips used in the test, of course, but training books and practice videos are available.

WHAT TO EXPECT IN THE PRACTICAL TEST

Once you have passed both parts of your Theory Test, you can apply for the Practical Test. The Practical Test is all about making sure that those who pass are competent and safe in the basic skills of driving.

> **The requirements for passing your test are a combination of practical skills and mental understanding. The open road can be a risky environment, and your test result will show whether you're ready to go out there alone or whether you need a little more practice first.**

> **You'll be asked to sign a declaration that the insurance of your car is in order. Without this, the test can't proceed.**

The paperwork

You'll need to have with you:

- your signed provisional driving licence (both parts if you've got a photo licence)
- your Theory Test pass certificate
- additional photographic identity (ID), if your licence doesn't have a photo
- your completed Driver's Record (if you have one) signed by your instructor.

If your licence does not show your photograph, you must also bring with you a form of photographic identification, such as a current signed passport. A card such as a workplace identity card, student union membership card or school bus pass is acceptable, but only if it is signed and has a photograph. For the full list of acceptable photographic ID, see the Driving Standards Agency website www.dsa.gov.uk

Eyesight test

Your driving test begins with an eyesight test. You have to be able to read a normal number plate at a minimum distance of 20.5 metres (about 67.5 feet). If you fail the eyesight test your driving test will stop at that point and you will have failed.

Vehicle safety checks

You will have to answer two vehicle safety check questions. The questions fall into three categories:

- identify
- tell me how you would check…
- show me how you would check…

These questions are designed to make sure that you know how to check that your vehicle is safe to drive.

> **If you turn up for your test in an unsuitable vehicle, you will forfeit your test fee.**

Although some checks may require you to identify where fluid levels should be checked you will not be asked to touch a hot engine or physically check fluid levels. You may refer to vehicle information systems (if fitted) when answering questions on fluid levels and tyre pressures.

All vehicles differ slightly so it is important that you get to know all the safety systems and engine layout in the vehicle in which you plan to take your practical test.

> **Don't worry about making a few mistakes. You can still pass your test as long as they are only minor driving faults.**

The Driving Test

During the test you will be expected to drive for about 40 minutes along normal roads following the directions of the examiner. The roads are selected so as to provide a range of different conditions and road situations and a varied density of traffic.

Your examiner will select suitable areas for you to carry out the set exercises. He or she will tell you to pull up and stop, then he or she will explain the exercise to you before you do it:

- you *may* or *may not* be asked to perform an emergency stop
- you *will* be asked to perform two reversing exercises selected by the examiner from:
  reversing round a corner;
  reverse parking (behind a parked car, or into a marked bay);
  turning in the road

Throughout the test, the examiner will be assessing:

- whether you are competent at controlling the car
- whether you are making normal progress for the roads you are on
- how you react to any hazards that occur in the course of the test

- whether you are noticing all traffic signs and signals and road markings, and reacting to them in the correct manner.

In order to pass the driving test, you must drive

- without committing any serious fault
  or
- without committing more than 15 driving errors of a less serious nature.

If, during the test, you do not understand what the examiner says to you, ask him or her to repeat the instruction.

If you are faced with an unusually difficult or hazardous situation in the course of your test that results in you making a driving fault, the examiner will take the circumstances into account when marking you for that part of the test.

How to prepare for the Practical Test

Be sure that you are ready to take the test. This is where choosing a reliable qualified driving instructor is vital.

> **Driving test standards are monitored so that whatever examiner you get, or whatever test centre you go to, you should get the same result. You might find a senior officer in the car as well as the examiner; he or she is not watching you, but checking that the examiner is doing his or her job properly.**

Record of your driving progress

Take this card with you to every lesson

Pupil Name

Provisional Licence No

AA Pupil No

Eyesight Checked - Date

First Driving Lesson Date

Instructor Name

Instructor No

No of hours tuition with an ADI

No of hours tuition without an ADI

Theory Test - Date Passed

Practical Test Date

Contact Tel No 1

Contact Tel No 2

How to use this record

You need to take this record with you each time you have a lesson and when you take the practical test. Your instructor will complete the record after each lesson and provide feedback on your progress. You are not ready to take your practical test until you reach stage 5 in all topics.

1 = introduced   **2** = under full instruction   **3** = prompted   **4** = seldom prompted   **5** = independent

Cockpit Checks
1 2 3 4 5   Date_____ Initials_____ Instructor No _____

Safety Check
1 2 3 4 5   Date_____ Initials_____ Instructor No _____

Controls & Instruments
1 2 3 4 5   Date_____ Initials_____ Instructor No _____

Moving Away & Stopping
1 2 3 4 5   Date_____ Initials_____ Instructor No _____

Other Traffic
1 2 3 4 5   Date_____ Initial_____ Instructor No _____

Junctions
1 2 3 4 5   Date_____ Ini_____ Instructor No _____

Roundabouts
1 2 3 4 5   Date_____ Instructor No _____

Pedestrian Crossings
1 2 3 4 5   Date_____ Instructor N_____

Dual Carriageways
_ 3 4 5   Date_____ Instruct_____

You should feel:

- confident about driving in all conditions
- confident that you know *The Highway Code*
- confident that you can make decisions on your own about how to cope with hazards, without having to wait for your instructor to tell you what to do.

Driver's Record

Completing a Driver's Record (see previous page) with your instructor should help you feel confident that you're ready to take your driving test.

The Driver's Record comes in two parts – one part for you and one for your instructor. It lists all the skills you need to master to become a safe driver and charts your progress in acquiring each of these skills through the following five levels:

1 Introduced
2 Under full instruction
3 Prompted
4 Seldom prompted
5 Independent

Take your copy of the Driver's Record to each lesson for your instructor to complete. Using the Private Practice sheet you can keep a record of driving experience gained when you are out driving with a friend or relative.

When your instructor has completed all the boxes in your Driver's Record you are ready to take your test. Remember to take the completed Driver's Record along with you to the driving test centre.

AFTER THE TEST

If you passed your test you'll be given a pass certificate, and a copy of the examiner's report showing any minor faults you made during your test. You'll find it useful to know where your minor weaknesses lie, so that you can concentrate on improving those aspects of your driving in the future.

With a full driving licence you are allowed to drive vehicles weighing up to 3.5 tonnes, use motorways for the first time, drive anywhere in the European Union and in many other countries worldwide, and to tow a small trailer. You are also on your own dealing with whatever circumstances may arise: fog, snow, ice, other drivers' mistakes. It is a huge responsibility.

Driving on a motorway for the first time can be a daunting experience. A good driving school will offer you the option of a post-test motorway lesson with your own instructor, and it makes sense to take advantage of this.

If you failed you will naturally be disappointed, but it's not the end of the world – many people don't pass their first test, but then sail through a second or third, having built on the experience of what it's like to take a test. The examiner will give you a test report form which is a record of all the skills assessed during the test, identifying any areas of weakness. He or she will also provide feedback in spoken form, and will explain to you what aspects of your driving are still in need of improvement.

Contents PAGE

Introduction

The Highway Code is essential reading for everyone. Its rules apply to all road users: pedestrians, horse riders and cyclists, as well as motorcyclists and drivers.

Many of the rules in the Code are legal requirements, and if you disobey these rules you are committing a criminal offence. You may be fined, given penalty points on your licence or be disqualified from driving. In the most serious cases you may be sent to prison. Such rules are identified by the use of the words **MUST/MUST NOT**. In addition, the rule includes an abbreviated reference to the legislation which creates the offence. An explanation of the abbreviations is in Annexe 4: The road user and the law.

Although failure to comply with the other rules of the Code will not, in itself, cause a person to be prosecuted, *The Highway Code* may be used in evidence in any court proceedings under the Traffic Acts to establish liability.

Knowing and applying the rules contained in *The Highway Code* could significantly reduce road accident casualties. Cutting the number of deaths and injuries that occur on our roads every day is a responsibility we all share. *The Highway Code* can help us discharge that responsibility.

Rules for pedestrians

General guidance

1. Pavements or footpaths should be used if provided. Where possible, avoid walking next to the kerb with your back to the traffic. If you have to step into the road, look both ways first.

2. If there is no pavement or footpath, walk on the right-hand side of the road so that you can see oncoming traffic. You should take extra care and
- be prepared to walk in single file, especially on narrow roads or in poor light
- keep close to the side of the road.

It may be safer to cross the road well before a sharp right-hand bend (so that oncoming traffic has a better chance of seeing you). Cross back after the bend.

3. Help other road users to see you. Wear or carry something light coloured, bright or fluorescent in poor daylight conditions. When it is dark, use reflective materials (e.g. armbands, sashes, waistcoats and jackets), which can be seen, by drivers using headlights, up to three times as far away as non-reflective materials.

Be seen in the dark; wear something reflective

4. Young children should not be out alone on the pavement or road (see Rule 7). When taking children out, walk between them and the traffic and hold their hands firmly. Strap very young children into pushchairs or use reins.

5. Organised walks. Groups of people should use a path if available; if one is not, they should keep to the left. Look-outs should be positioned at the front and back of the group, and they should wear fluorescent clothes in daylight and reflective clothes in the dark. At night, the look-out in front should carry a white light and the one at the back a red light. People on the outside of large groups should also carry lights and wear reflective clothing.

6. Motorways. You **MUST NOT** walk on motorways or slip roads except in an emergency (see Rule 249).

Laws RTRA sect 17, MT(E&W)R 1982 as amended & MT(S)R regs 2 & 13

Crossing the road

7. The Green Cross Code. The advice given below on crossing the road is for all pedestrians. Children should be taught the Code and should not be allowed out alone until they can understand and use it properly. The age when they can do this is different for each child. Many children cannot judge how fast vehicles are going or how far away they are. Children learn by example, so parents and carers should always use the Code in full when out with their children. They are responsible for deciding at what age children can use it safely by themselves.

a. First find a safe place to cross. It is safer to cross using a subway, a footbridge, an island, a zebra, pelican, toucan or puffin crossing, or where there is a crossing point controlled by a police officer, a school crossing patrol or a traffic warden. Where there is a crossing nearby, use it. Otherwise choose a place where you can see clearly in all directions. Try to avoid crossing between parked cars (see Rule 14) and on blind bends and brows of hills. Move to a space where drivers can see you clearly.

b. Stop just before you get to the kerb, where you can see if anything is coming. Do not get too close to the traffic. If there is no pavement, keep back from the edge of the road but make sure you can still see approaching traffic.

c. Look all around for traffic and listen. Traffic could come from any direction. Listen as well, because you can sometimes hear traffic before you see it.

d. If traffic is coming, let it pass. Look all around again and listen. Do not cross until there is a safe gap in the traffic and you are certain that there is plenty of time. Remember, even if traffic is a long way off, it may be approaching very quickly.

e. When it is safe, go straight across the road – do not run. Keep looking and listening for traffic while you cross, in case there is any traffic you did not see, or in case other traffic appears suddenly.

8. At a junction. When crossing the road, look out for traffic turning into the road, especially from behind you.

9. Pedestrian Safety Barriers. Where there are barriers, cross the road only at the gaps provided for pedestrians. Do not climb over the barriers or walk between them and the road.

10. Tactile paving. Small raised studs which can be felt underfoot may be used to advise blind or partially-sighted people that they are approaching a crossing point with a dropped kerb.

11. One-way streets. Check which way the traffic is moving. Do not cross until it is safe to do so without stopping. Bus and cycle lanes may operate in the opposite direction to the rest of the traffic.

12. Bus and cycle lanes. Take care when crossing these lanes as traffic may be moving faster than in the other lanes, or against the flow of traffic.

13. Routes shared with cyclists. Cycle tracks may run alongside footpaths, with a dividing line segregating the two. Keep to the section for pedestrians. Take extra care where cyclists and pedestrians share the same path without separation (see Rule 48).

14. Parked vehicles. If you have to cross between parked vehicles, use the outside edges of the vehicles as if they were the kerb. Stop there and make sure you can see all around and that the traffic can see you. Never cross the road in front of, or behind, any vehicle with its engine running, especially a large vehicle, as the driver may not be able to see you.

15. Reversing vehicles. Never cross behind a vehicle which is reversing, showing white reversing lights or sounding a warning.

16. Moving vehicles. You **MUST NOT** get on to or hold on to a moving vehicle.

Law RTA 1988 sect 26

17. At night. Wear something reflective to make it easier for others to see you (see Rule 3). If there is no pedestrian crossing nearby, cross the road near a street light so that traffic can see you more easily.

Crossings

18. At all crossings. When using any type of crossing you should

- always check that the traffic has stopped before you start to cross or push a pram on to a crossing
- always cross between the studs or over the zebra markings. Do not cross at the side of the crossing or on the zig-zag lines, as it can be dangerous.

You **MUST NOT** loiter on zebra, pelican or puffin crossings.

Laws ZPPPCRGD reg 19 & RTRA sect 25(5)

19. Zebra crossings. Give traffic plenty of time to see you and to stop before you start to cross. Vehicles will need more time when the road is slippery. Remember that traffic does not have to stop until someone has moved on to the crossing. Wait until traffic has stopped from both directions or the road is clear before crossing. Keep looking both ways, and listening, in case a driver or rider has not seen you and attempts to overtake a vehicle that has stopped.

20. Where there is an island in the middle of a zebra crossing, wait on the island and follow Rule 19 before you cross the second half of the road – it is a separate crossing.

21. At traffic lights. There may be special signals for pedestrians. You should only start to cross the road when the green figure shows. If you have started to cross the road and the green figure goes out, you should still have time to reach the other side, but do not delay. If no pedestrian signals have been provided, watch carefully and do not cross until the traffic lights are red and the traffic has stopped. Keep looking and check for traffic that may be turning the corner. Remember that traffic lights may let traffic move in some lanes while traffic in other lanes has stopped.

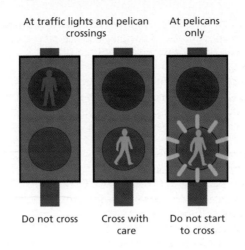

At traffic lights and pelican crossings | At pelicans only

Do not cross | Cross with care | Do not start to cross

Pedestrian signals at traffic lights and pelican crossings

22. Pelican crossings. These are signal-controlled crossings operated by pedestrians. Push the control button to activate the traffic signals. When the red figure shows, do not cross. When a steady green figure shows, check the traffic has stopped then cross with care. When the green figure begins to flash you should not start to cross. If you have already started you should have time to finish crossing safely.

23. At some pelican crossings there is a bleeping sound to indicate to blind or partially-sighted people when the steady green figure is showing, and there may be a tactile signal to help deafblind people.

24. When the road is congested, traffic on your side of the road may be forced to stop even though their lights are green. Traffic may still be moving on the other side of the road, so press the button and wait for the signal to cross.

25. Puffin and toucan crossings. These differ from pelican crossings as there is no flashing green figure phase. On puffin crossings the red and green figures are above the control box on your side of the road. Press the button and wait for the green figure to show. On toucan crossings cyclists are permitted to ride across the road (see Rule 65).

26. 'Staggered' pelican or puffin crossings. When the crossings on each side of the central refuge are not in line they are two separate crossings. On reaching the central island press the button again and wait for a steady green figure.

27. Crossings controlled by an authorised person. Do not cross the road unless you are signalled to do so by a police officer, traffic warden or school crossing patrol. Always cross in front of them.

28. Where there are no controlled crossing points available it is advisable to cross where there is an island in the middle of the road. Use the Green Cross Code to cross to the island and then stop and use it again to cross the second half of the road.

Situations needing extra care

29. Emergency vehicles. If an ambulance, fire engine, police or other emergency vehicle approaches using flashing blue lights, headlights and/or sirens, keep off the road.

30. Buses. Get on or off a bus only when it has stopped to allow you to do so. Watch out for cyclists when you are getting off. Never cross the road directly behind or in front of a bus; wait until it has moved off and you can see clearly in both directions.

31. Tramways. These may run through pedestrian areas. Their path will be marked out by shallow kerbs, changes in the paving or other road surface, white lines or yellow dots. Cross at designated crossings where provided. Flashing amber lights may warn you that a tram is approaching. Elsewhere look both ways along the track before crossing. Do not walk along the track. Trams move quickly and silently and cannot steer to avoid you.

32. Railway level crossings. Do not cross if the red lights show, an alarm is sounding or the barriers are being lowered. The tone of the alarm will change if another train is approaching. If there are no lights, alarms or barriers, stop, look both ways and listen before crossing.

33. Street and pavement repairs. A pavement may be closed temporarily because it is not safe to use. Take extra care if you are directed to walk in or to cross the road.

Rules about animals

Horseriders

34. Safety equipment. Children under the age of 14 **MUST** wear a helmet which complies with the Regulations. It **MUST** be fastened securely. Other riders should also follow this advice.
Law H(PHYR)R

35. Other clothing. You should wear
- boots or shoes with hard soles and heels
- light-coloured or fluorescent clothing in daylight
- reflective clothing if you have to ride at night or in poor visibility.

36. At night. It is safer not to ride on the road at night or in poor visibility, but if you do, make sure your horse has reflective bands above the fetlock joints. Carry a light which shows white to the front and red to the rear.

Riding

37. Before you take a horse on to a road, you should
- ensure all tack fits well and is in good condition
- make sure you can control the horse.

Always ride with other, less nervous horses if you think that your horse will be nervous of traffic. Never ride a horse without a saddle or bridle.

38. Before riding off or turning, look behind you to make sure it is safe, then give a clear arm signal.

39. When riding on the road you should
- keep to the left
- keep both hands on the reins unless you are signalling
- keep both feet in the stirrups
- not carry another person
- not carry anything which might affect your balance or get tangled up with the reins
- keep a horse you are leading to your left
- move in the direction of the traffic flow in a one-way street
- never ride more than two abreast, and ride in single file where the road narrows or on the approach to a bend.

40. You **MUST NOT** take a horse on to a footpath, pavement or cycle track. Use a bridleway where possible.
Laws HA 1835 sect 72 & R(S)A sect 129(5)

41. Avoid roundabouts wherever possible. If you use them you should
- keep to the left and watch out for vehicles crossing your path to leave or join the roundabout
- signal right when riding across exits to show you are not leaving
- signal left just before you leave the roundabout.

Other animals

42. Dogs. Do not let a dog out on the road on its own. Keep it on a short lead when walking on the pavement, road or path shared with cyclists.

43. When in a vehicle make sure dogs or other animals are suitably restrained so they cannot distract you while you are driving or injure you if you stop quickly.

44. Animals being herded. These should be kept under control at all times. You should, if possible, send another person along the road in front to warn other road users, especially at a bend or the brow of a hill. It is safer not to move animals after dark, but if you do, then wear reflective clothing and ensure that lights are carried (white at the front and red at the rear of the herd).

Rules for cyclists

These rules are in addition to those in the following sections, which apply to all vehicles (except the motorway section). See also Annexe 1: Choosing and maintaining your bicycle.

45. Clothing. You should wear
- a cycle helmet which conforms to current regulations
- appropriate clothes for cycling. Avoid clothes which may get tangled in the chain, or in a wheel or may obscure your lights
- light-coloured or fluorescent clothing which helps other road users to see you in daylight and poor light
- reflective clothing and/or accessories (belt, arm or ankle bands) in the dark.

Help yourself to be seen

46. At night your cycle **MUST** have front and rear lights lit. It **MUST** also be fitted with a red rear reflector (and amber pedal reflectors, if manufactured after 1/10/85). White front reflectors and spoke reflectors will also help you to be seen.
Law RVLR regs 18 & 24

When cycling
47. Use cycle routes when practicable. They can make your journey safer.

48. Cycle Tracks. These are normally located away from the road, but may occasionally be found alongside footpaths or pavements. Cyclists and pedestrians may be segregated or they may share the same space (unsegregated). When using segregated tracks you **MUST** keep to the side intended for cyclists. Take care when passing pedestrians, especially children, elderly or disabled people, and allow them plenty of room. Always be prepared to slow down and stop if necessary.

Law HA 1835 sect 72

49. Cycle Lanes. These are marked by a white line (which may be broken) along the carriageway (see Rule 119). Keep within the lane wherever possible.

50. You **MUST** obey all traffic signs and traffic light signals.

Laws RTA 1988 sect 36, TSRGD reg 10(1)

51. You should
- keep both hands on the handlebars except when signalling or changing gear
- keep both feet on the pedals
- not ride more than two abreast
- ride in single file on narrow or busy roads
- not ride close behind another vehicle
- not carry anything which will affect your balance or may get tangled up with your wheels or chain
- be considerate of other road users, particularly blind and partially-sighted pedestrians. Let them know you are there when necessary, for example by ringing your bell.

52. You should
- look all around before moving away from the kerb, turning or manoeuvring, to make sure it is safe to do so. Give a clear signal to show other road users what you intend to do (see Signals to other road users)
- look well ahead for obstructions in the road, such as drains, pot-holes and parked vehicles so that you do not have to swerve suddenly to avoid them. Leave plenty of room when passing parked vehicles and watch out for doors being opened into your path
- take extra care near road humps, narrowings and other traffic calming features.

53. You **MUST NOT**
- carry a passenger unless your cycle has been built or adapted to carry one
- hold on to a moving vehicle or trailer
- ride in a dangerous, careless or inconsiderate manner
- ride when under the influence of drink or drugs.

Law RTA 1988 sects 24, 26, 28, 29 & 30 as amended by RTA 1991

54. You **MUST NOT** cycle on a pavement. Do not leave your cycle where it would endanger or obstruct road users or pedestrians, for example, lying on the pavement. Use cycle parking facilities where provided.

Laws HA 1835 sect 72 & R(S)A sect 129

55. You **MUST NOT** cross the stop line when the traffic lights are red. Some junctions have an advanced stop line to enable you to position yourself ahead of other traffic (see Rule 154).

Laws RTA 1988 sect 36, TSRGD regs 10 & 36(1)

56. Bus Lanes. These may be used by cyclists only if the signs include a cycle symbol. Watch out for people getting on or off a bus. Be very careful when overtaking a bus or leaving a bus lane as you will be entering a busier traffic flow.

Road junctions

57. On the left. When approaching a junction on the left, watch out for vehicles turning in front of you, out of or into the side road. Do not ride on the inside of vehicles signalling or slowing down to turn left.

58. Pay particular attention to long vehicles which need a lot of room to manoeuvre at corners. They may have to move over to the right before turning left. Wait until they have completed the manoeuvre because the rear wheels come very close to the kerb while turning. Do not be tempted to ride in the space between them and the kerb.

59. On the right. If you are turning right, check the traffic to ensure it is safe, then signal and move to the centre of the road. Wait until there is a safe gap in the oncoming traffic before completing the turn. It may be safer to wait on the left until there is a safe gap or to dismount and push your cycle across the road.

60. Dual carriageways. Remember that traffic on most dual carriageways moves quickly. When crossing wait for a safe gap and cross each carriageway in turn. Take extra care when crossing slip roads.

Roundabouts

61. Full details about the correct procedure at roundabouts are contained in Rules 160–166. Roundabouts can be hazardous and should be approached with care.

62. You may feel safer either keeping to the left on the roundabout or dismounting and walking your cycle round on the pavement or verge. If you decide to keep to the left you should

- be aware that drivers may not easily see you
- take extra care when cycling across exits and you may need to signal right to show you are not leaving the roundabout
- watch out for vehicles crossing your path to leave or join the roundabout.

63. Give plenty of room to long vehicles on the roundabout as they need more space to manoeuvre. Do not ride in the space they need to get round the roundabout. It may be safer to wait until they have cleared the roundabout.

Crossing the road
64. Do not ride across a pelican, puffin or zebra crossing. Dismount and wheel your cycle across.

65. Toucan crossings. These are light-controlled crossings which allow cyclists and pedestrians to cross at the same time. They are push button operated. Pedestrians and cyclists will see the green signal together. Cyclists are permitted to ride across.

66. Cycle-only crossings. Cycle tracks on opposite sides of the road may be linked by signalled crossings. You may ride across but you **MUST NOT** cross until the green cycle symbol is showing.
Law TSRGD reg 36(1)

Rules for motorcyclists

These Rules are in addition to those in the following sections which apply to all vehicles. For motorcycle licence requirements see Annexe 2: Motorcycle licence requirements.

General

67. On all journeys, the rider and pillion passenger on a motorcycle, scooter or moped **MUST** wear a protective helmet. Helmets **MUST** comply with the Regulations and they **MUST** be fastened securely. It is also advisable to wear eye protectors, which **MUST** comply with the Regulations. Consider wearing ear protection. Strong boots, gloves and suitable clothing may help to protect you if you fall off.

Laws RTA 1988 sects 16 &17 & MC(PH)R as amended reg 4, & RTA sect 18 & MC(EP)R as amended reg 4

68. You **MUST NOT** carry more than one pillion passenger and he/she **MUST** sit astride the machine on a proper seat and should keep both feet on the footrests.

Law RTA 1988 sect 23

69. Daylight riding. Make yourself as visible as possible from the side as well as the front and rear. You could wear a white or brightly-coloured helmet. Wear fluorescent clothing or strips. Dipped headlights, even in good daylight, may also make you more conspicuous.

Make sure you can be seen

70. Riding in the dark. Wear reflective clothing or strips to improve your chances of being seen in the dark. These reflect light from the headlamps of other vehicles making you more visible from a long distance. See Rules 93–96 for lighting requirements.

71. Manoeuvring. You should be aware of what is behind and to the sides before manoeuvring. Look behind you; use mirrors if they are fitted. When overtaking traffic queues look out for pedestrians crossing between vehicles and vehicles emerging from junctions.
Remember: Observation – Signal – Manoeuvre.

Rules for drivers and motorcyclists

72. Vehicle condition. You **MUST** ensure your vehicle and trailer complies with the full requirements of the Road Vehicles (Construction and Use) Regulations and Road Vehicles Lighting Regulations. (See Annexe 6: Vehicle maintenance safety and security).

73. Before setting off. You should ensure that
- you have planned your route and allowed sufficient time
- clothing and footwear do not prevent you using the controls in the correct manner
- you know where all the controls are and how to use them before you need them. All vehicles are different; do not wait until it is too late to find out
- your mirrors and seat are adjusted correctly to ensure comfort, full control and maximum vision
- head restraints are properly adjusted to reduce the risk of neck injuries in the event of an accident
- you have sufficient fuel before commencing your journey, especially if it includes motorway driving. It can be dangerous to lose power when driving in traffic.

74. Vehicle towing and loading. As a driver
- you **MUST NOT** tow more than your licence permits you to
- you **MUST NOT** overload your vehicle or trailer. You should not

tow a weight greater than that recommended by the manufacturer of your vehicle

- you **MUST** secure your load and it **MUST NOT** stick out dangerously
- you should properly distribute the weight in your caravan or trailer with heavy items mainly over the axle(s) and ensure a downward load on the tow ball. Manufacturer's recommended weight and tow ball load should not be exceeded. This should avoid the possibility of swerving or snaking and going out of control. If this does happen, ease off the accelerator and reduce speed gently to regain control.

Law CUR reg 100, MV(DL)R reg 43

Seat Belts

75. You **MUST** wear a seat belt if one is available, unless you are exempt. Those exempt from the requirement include the holders of medical exemption certificates and people making local deliveries in a vehicle designed for the purpose.

Laws RTA 1988 sects 14 & 15, MV(WSB)R & MV(WSBCFS)R

Seat belt requirements

This table summarises the main legal requirements for wearing seat belts

	FRONT SEAT (all vehicles)	REAR SEAT (cars and small minibuses*)	WHOSE RESPONSIBILITY
DRIVER	**MUST** be worn if fitted		**DRIVER**
CHILD under 3 years of age	Appropriate child restraint **MUST** be worn	Appropriate child restraint **MUST** be worn *if available*	**DRIVER**
CHILD aged 3 to 11 and under 1.5 metres (about 5 feet) in height	Appropriate child restraint **MUST** be worn *if available.* If not, an adult seat belt **MUST** be worn	Appropriate child restraint **MUST** be worn *if available.* If not, an adult seat belt **MUST** be worn *if available*	**DRIVER**
CHILD aged 12 or 13 or younger child 1.5 metres or more in height	Adult seat belt **MUST** be worn *if available*	Adult seat belt **MUST** be worn *if available*	**DRIVER**
PASSENGER over the age of 14	**MUST** be worn *if available*	**MUST** be worn *if available*	**PASSENGER**

*Minibuses with an unladen weight of 2540kg or less

76. The driver **MUST** ensure that all children under 14 years of age wear seat belts or sit in an approved child restraint. This should be a baby seat, child seat, booster seat or booster cushion appropriate to the child's weight and size, fitted to the manufacturer's instructions.
Laws RTA 1988 sects 14 & 15, MV(WSB)R & MV(WSBCFS)R

Make sure children wear the correct restraint

77. You **MUST** wear seat belts in minibuses with an unladen weight of 2540 kg or less. You should wear them in large minibuses and coaches where available.
Laws RTA 1988 sects 14 & 15, MV(WSB)R & MV(WSBCFS)R

78. Children in cars. Drivers who are carrying children in cars should ensure that
- children do not sit behind the rear seats in an estate car or hatchback, unless a special child seat has been fitted
- the child safety door locks, where fitted, are used when children are in the car
- children are kept under control
- a rear-facing baby seat is **NEVER** fitted into a seat protected by an airbag.

Fitness to drive
79. Make sure that you are fit to drive. You **MUST** report to the Driver and Vehicle Licensing Agency (DVLA) any health condition likely to affect your driving.
Law RTA 1988 sect 94

80. Driving when you are tired greatly increases your accident risk. To minimise this risk
- make sure you are fit to drive. Do not undertake a long journey (longer than an hour) if you feel tired
- avoid undertaking long journeys between midnight and 6am, when natural alertness is at a minimum

- plan your journey to take sufficient breaks. A minimum break of at least 15 minutes after every two hours of driving is recommended
- if you feel at all sleepy, stop in a safe place. Do not stop on the hard shoulder of a motorway
- the most effective ways to counter sleepiness are to take a short nap (up to 15 minutes) or drink, for example, two cups of strong coffee. Fresh air, exercise or turning up the radio may help for a short time, but are *not* as effective.

81. Vision. You **MUST** be able to read a vehicle number plate from a distance of 20.5 metres (67 feet – about five car lengths) in good daylight. From September 2001, you **MUST** be able to read a new style number plate from a distance of 20 metres (66 feet). If you need to wear glasses (or contact lenses) to do this, you **MUST** wear them at all times whilst driving. The police have the power to require a driver, at any time, to undertake an eyesight test in good daylight.
Laws RTA 1988 sect 96 & MV(DL)R reg 40 & sch 8

82. At night or in poor visibility, do not use tinted glasses, lenses or visors or anything that restricts vision.

Alcohol and drugs
83. Do not drink and drive as it will seriously affect your judgement and abilities. You **MUST NOT** drive with a breath alcohol level higher than 35 µg /100 ml or a blood alcohol level of more than 80 mg/100 ml. Alcohol will
- give a false sense of confidence
- reduce co-ordination and slow down reactions
- affect judgement of speed, distance and risk
- reduce your driving ability, even if you are below the legal limit
- take time to leave your body; you may be unfit to drive in the evening after drinking at lunchtime, or in the morning after drinking the previous evening. If you are going to drink, arrange another means of transport.
Law RTA 1988 sects 4, 5 & 11(2)

84. You **MUST NOT** drive under the influence of drugs or medicine. Check the instructions or ask your doctor or pharmacist. Using illegal drugs is highly dangerous. Never take them before driving; the effects are unpredictable, but can be even more severe than alcohol and may result in fatal or serious road accidents.
Law RTA 1988 sect 4

General rules, techniques and advice for all drivers and riders

This section should be read by all drivers, motorcyclists and cyclists. The rules in *The Highway Code* do not give you the right of way in any circumstance, but they advise you when you should give way to others. Always give way if it can help to avoid an accident.

Signals

85. Signals warn and inform other road users, including pedestrians (see Signals to other road users), of your intended actions.

You should
- give clear signals in plenty of time, having checked it is not misleading to signal at that time
- use them, if necessary, before changing course or direction, stopping or moving off
- cancel them after use
- make sure your signals will not confuse others. If, for instance, you want to stop after a side road, do not signal until you are passing the road. If you signal earlier it may give the impression that you intend to turn into the road. Your brake lights will warn traffic behind you that you are slowing down
- use an arm signal to emphasise or reinforce your signal if necessary. Remember that signalling does not give you priority.

86. You should also
- watch out for signals given by other road users and proceed only when you are satisfied that it is safe
- be aware that an indicator on another vehicle may not have been cancelled.

87. You **MUST** obey signals given by police officers and traffic wardens (see Signals by authorised persons) and signs used by school crossing patrols.
Laws RTRA sect 28, RTA 1988 sect 35 and FTWO art 3

Traffic light signals and traffic signs
88. You **MUST** obey all traffic light signals (see Light signals controlling traffic) and traffic signs giving orders, including temporary signals and signs (see Traffic Signs and Road works signs). Make sure you know, understand and act on all other traffic and information signs and road

markings (see Traffic signs and Road markings).
Laws RTA 1988 sect 36, TSRGD regs 10,15,16,25,26,27,28,29,36,38 & 40

89. Police stopping procedures. If the police want to stop your vehicle they will, where possible, attract your attention by
- flashing blue lights or headlights or sounding their siren or horn
- directing you to pull over to the side by pointing and/or using the left indicator.

You **MUST** then pull over and stop as soon as it is safe to do so. Then switch off your engine.
Law RTA 1988 sect 163

90. Flashing headlights. Only flash your headlights to let other road users know that you are there. Do not flash your headlights in an attempt to intimidate other road users.

91. If another driver flashes his headlights never assume that it is a signal to go. Use your own judgement and proceed carefully.

92. The horn. Use only while your vehicle is moving and you need to warn other road users of your presence. Never sound your horn aggressively. You **MUST NOT** use your horn
- while stationary on the road
- when driving in a built-up area between the hours of 11.30 pm and 7.00 am

except when another vehicle poses a danger.
Law CUR reg 99

Lighting requirements

93. You **MUST**
- use headlights at night, except on restricted roads (those with street lights not more than 185 metres (600 feet) apart and which are generally subject to a speed limit of 30 mph)
- use headlights when visibility is seriously reduced (see Rule 201)
- ensure all sidelights and rear registration plate lights are lit at night.

Laws RVLR regs 24 & 25 & RV(R&L)R reg 19

94. You **MUST NOT**
- use any lights in a way which would dazzle or cause discomfort to other road users
- use front or rear fog lights unless visibility is seriously reduced. You **MUST** switch them off when visibility improves to avoid dazzling other road users.

Law RVLR reg 27

95. You should also
- use dipped headlights, or dim-dip if fitted, at night in built-up areas and in dull daytime weather, to ensure that you can be seen
- keep your headlights dipped when overtaking until you are level with the other vehicle and then change to main beam if necessary, unless this would dazzle oncoming traffic
- slow down, and if necessary stop, if you are dazzled by oncoming headlights.

96. Hazard warning lights. These may be used when your vehicle is stationary, to warn that it is temporarily obstructing traffic. Never use them as an excuse for dangerous or illegal parking. You **MUST NOT** use hazard warning lights whilst driving unless you are on a motorway or unrestricted dual carriageway and you need to warn drivers behind you of a hazard or obstruction ahead. Only use them for long enough to ensure that your warning has been observed.
Law RVLR reg 27

Control of the vehicle
Braking

97. In normal circumstances. The safest way to brake is to do so early and lightly. Brake more firmly as you begin to stop. Ease the pressure off just before the vehicle comes to rest to avoid a jerky stop.

98. In an emergency. Brake immediately. Try to avoid braking so harshly that you lock your wheels. Locked wheels can lead to skidding.

99. Skids. Skidding is caused by the driver braking, accelerating or steering too harshly or driving too fast for the road conditions. If skidding occurs, ease off the brake or accelerator and try to steer smoothly in the direction of the skid. For example, if the rear of the vehicle skids to the right, steer quickly and smoothly to the right to recover.

Rear of car skids to the right

Driver steers to the right

100. ABS. The presence of an anti-lock braking system should not cause you to alter the way you brake from that indicated in Rule 97. However in the case of an emergency, apply the footbrake rapidly and firmly; do not release the pressure until the vehicle has slowed to the desired speed. The ABS should ensure that steering control will be retained.

101. Brakes affected by water. If you have driven through deep water your brakes may be less effective. Test them at the first safe opportunity by pushing gently on the brake pedal to make sure that they work. If they are not fully effective, gently apply light pressure while driving slowly. This will help to dry them out.

102. Coasting. This term describes a vehicle travelling in neutral or with the clutch pressed down. Do not coast, whatever the driving conditions. It reduces driver control because
- engine braking is eliminated
- vehicle speed downhill will increase quickly
- increased use of the footbrake can reduce its effectiveness
- steering response will be affected particularly on bends and corners
- it may be more difficult to select the appropriate gear when needed.

Speed limits
103. You **MUST NOT** exceed the maximum speed limits for the road and for your vehicle (see the table on the next page). Street lights usually mean that there is a 30 mph speed limit unless there are signs showing another limit.
Law RTRA sects 81,86,89 & sch 6

Speed Limits

Type of vehicle	Built-up areas* MPH	Elsewhere Single carriageways MPH	Elsewhere Dual carriageways MPH	Motorways MPH
Cars & motorcycles (including car derived vans up to 2 tonnes maximum laden weight)	30	60	70	70
Cars towing caravans or trailers (including car derived vans and motorcycles)	30	50	60	60
Buses & coaches (not exceeding 12 metres in overall length)	30	50	60	70
Goods vehicles (not exceeding 7.5 tonnes maximum laden wieght)	30	50	60	70†
Goods vehicles (exceeding 7.5 tonnes maximum laden wieght)	30	40	50	60

These are the national speed limits and apply to all roads unless signs show otherwise.
*The 30mph limit applies to all traffic on all roads in England and Wales (only class C and unclassified roads in Scotland) with street lighting unless signs show otherwise.
†60 if articulated or towing a trailer

104. The speed limit is the absolute maximum and does not mean it is safe to drive at that speed irrespective of conditions. Driving at speeds too fast for the road and traffic conditions can be dangerous. You should always reduce your speed when

- the road layout or condition presents hazards, such as bends
- sharing the road with pedestrians and cyclists, particularly children, and motorcyclists
- weather conditions make it safer to do so
- driving at night as it is harder to see other road users.

Stopping distances

105. Drive at a speed that will allow you to stop well within the distance you can see to be clear. You should

- leave enough space between you and the vehicle in front so that you can pull up safely if it suddenly slows down or stops. The safe rule is never to get closer than the overall stopping distance (see Typical Stopping Distances diagram, on the next page)
- allow at least a two-second gap between you and the vehicle in front on roads carrying fast traffic. The gap should be at least doubled on wet roads and increased still further on icy roads
- remember, large vehicles and motorcycles need a greater distance to stop.

Use a fixed point to help measure a two-second gap

Typical stopping Distances

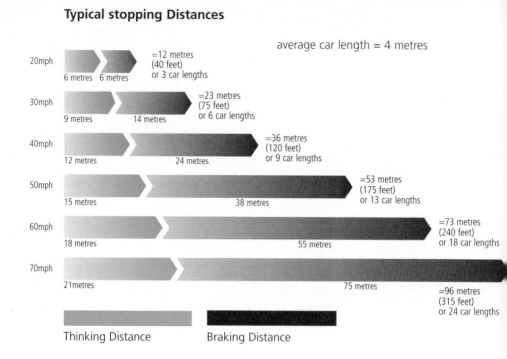

average car length = 4 metres

20mph — 6 metres, 6 metres =12 metres (40 feet) or 3 car lengths

30mph — 9 metres, 14 metres =23 metres (75 feet) or 6 car lengths

40mph — 12 metres, 24 metres =36 metres (120 feet) or 9 car lengths

50mph — 15 metres, 38 metres =53 metres (175 feet) or 13 car lengths

60mph — 18 metres, 55 metres =73 metres (240 feet) or 18 car lengths

70mph — 21metres, 75 metres =96 metres (315 feet) or 24 car lengths

Thinking Distance   Braking Distance

Lines and lane markings on the road

Diagrams of all lines shown in Road markings.

106. A broken white line. This marks the centre of the road. When this line lengthens and the gaps shorten, it means that there is a hazard ahead. Do not cross it unless you can see the road is clear well ahead and wish to overtake or turn off.

107. Double white lines where the line nearest to you is broken. This means you may cross the lines to overtake if it is safe, provided you can complete the manoeuvre before reaching a solid white line on your side. White arrows on the road indicate when you need to get back onto your side of the road.

108. Double white lines where the line nearest you is solid. This means you **MUST NOT** cross or straddle it unless it is safe and you need to enter adjoining premises or a side road. You may cross the line if necessary to pass a stationary vehicle, or overtake a pedal cycle, horse or road maintenance vehicle, if they are travelling at 10 mph or less.

Laws RTA sect 36 & TSRGD regs 10 & 26

109. Areas of white diagonal stripes or chevrons painted on the road. These are to separate traffic lanes or to protect traffic turning right.

- If the area is bordered by a broken white line, you should not enter the area unless it is necessary and you can see that it is safe to do so.
- If the area is marked with diagonal stripes and bordered by solid white lines, you should not enter it except in an emergency.
- If the area is marked with chevrons and bordered by solid white lines you **MUST NOT** enter it except in an emergency.

Laws MT(E&W)R regs 5,9,10 & 16, MT(S)R regs 4,8,9 & 14, RTA sect 36 & TSRGD 10(1)

110. Lane dividers. These are short broken white lines which are used on wide carriageways to divide them into lanes. You should keep between them.

111. Reflective road studs may be used with white lines.

- White studs mark the lanes or the middle of the road.
- Red studs mark the left edge of the road.
- Amber studs mark the central reservation of a dual carriageway or motorway.
- Green studs mark the edge of the main carriageway at lay-bys, side roads and slip roads.

Multi-lane carriageways

Lane discipline

112. If you need to change lane, first use your mirrors and check your blind spots (the areas you are unable to see in the mirrors) to make sure you will not force another driver or rider to swerve or slow down. When it is safe to do so, signal to indicate your intentions to other road users and when clear move over.

113. You should follow the signs and road markings and get into lane as directed. In congested road conditions do not change lanes unnecessarily.

Single carriageway

114. Where a single carriageway has three lanes and the road markings or signs do not give priority to traffic in either direction

- use the middle lane only for overtaking or turning right. Remember, you have no more right to use the middle lane than a driver coming from the opposite direction
- do not use the right-hand lane.

115. Where a single carriageway has four or more lanes, use only the lanes that signs or markings indicate.

Dual carriageways

116. On a two-lane dual carriageway you should stay in the left-hand lane. Use the right-hand lane for overtaking or turning right. If you use it for overtaking move back to the left-hand lane when it is safe to do so.

117. On a three-lane dual carriageway, you may use the middle lane or the right-hand lane to overtake but return to the middle and then the left-hand lane when it is safe.

118. Climbing and crawler lanes. These are provided on some hills. Use this lane if you are driving a slow moving vehicle or if there are vehicles behind you wishing to overtake.

119. Cycle lanes. These are shown by road markings and signs. You **MUST NOT** drive or park in a cycle lane marked by a solid white line during its times of operation. Do not drive or park in a cycle lane marked by a broken white line unless it is unavoidable. You **MUST NOT** park in any cycle lane whilst waiting restrictions apply.

Law RTRA sects 5 & 8

120. Bus and tram lanes. These are shown by road markings and signs. You **MUST NOT** drive or stop in a tram lane or in a bus lane during its period of operation unless the signs indicate you may do so.
Law RTRA sects 5 & 8

121. One-way streets. Traffic **MUST** travel in the direction indicated by signs. Buses and/or cycles may have a contraflow lane. Choose the correct lane for your exit as soon as you can. Do not change lanes suddenly. Unless road signs or markings indicate otherwise, you should use
- the left-hand lane when going left
- the right-hand lane when going right
- the most appropriate lane when going straight ahead.
Remember – traffic could be passing on both sides.
Laws RTA 1988 sect 36 & RTRA sects 5 & 8

General advice

122. You MUST NOT
- drive dangerously
- drive without due care and attention
- drive without reasonable consideration for other road users.
Law RTA 1988 sects 2 & 3 as amended by RTA 1991

123. You **MUST NOT** drive on or over a pavement, footpath or bridleway except to gain lawful access to property.
Laws HA 1835 sect 72 & RTA sect 34

124. Adapt your driving to the appropriate type and condition of road you are on. In particular
- do not treat speed limits as a target. It is often not appropriate or safe to drive at the maximum speed limit
- take the road and traffic conditions into account. Be prepared for unexpected or difficult situations, for example, the road being blocked beyond a blind bend. Be prepared to adjust your speed as a precaution
- where there are junctions, be prepared for vehicles emerging
- in side roads and country lanes look out for unmarked junctions where nobody has priority
- try to anticipate what pedestrians and cyclists might do.
  If pedestrians, particularly children, are looking the other way, they may step out into the road without seeing you.

125. Be considerate. Be careful of and considerate towards other road users. You should
- try to be understanding if other drivers cause problems; they may be inexperienced or not know the area well
- be patient; remember that anyone can make a mistake
- not allow yourself to become agitated or involved if someone is behaving badly on the road. This will only make the situation worse. Pull over, calm down and, when you feel relaxed, continue your journey
- slow down and hold back if a vehicle pulls out into your path at a junction. Allow it to get clear. Do not over-react by driving too close behind it.

126. Safe driving needs concentration. Avoid distractions when driving such as
- loud music (this may mask other sounds)
- trying to read maps
- inserting a cassette or CD or tuning a radio
- arguing with your passengers or other road users
- eating and drinking.

Mobile phones and in-car technology

127. You **MUST** exercise proper control of your vehicle at all times. You **MUST NOT** use a hand-held mobile phone, or similiar device, when driving or when supervising a learner driver, except to call 999 or 112 in a genuine emergency when it is unsafe or impractical to stop. Never use a hand-held microphone when driving. Using hands-free equipment is also likely to distract your attention from the road. It is far safer not to use any telephone while you are driving – find a safe place to stop first.
Laws RTA 1988 sects 2 & 3 & CUR regs 104 & 110

128. There is a danger of driver distraction being caused by in-vehicle systems such as route guidance and navigation systems, congestion warning systems, PCs, multi-media, etc. Do not operate, adjust or view any such system if it will distract your attention while you are driving; you **MUST** exercise proper control of your vehicle at all times. If necessary find a safe place to stop first.
Laws RTA 1988 sects 2 & 3 & CUR reg 104

In slow moving traffic

129. You should
- reduce the distance between you and the vehicle ahead to maintain traffic flow
- never get so close to the vehicle in front that you cannot stop safely

- leave enough space to be able to manoeuvre if the vehicle in front breaks down or an emergency vehicle needs to get past
- not change lanes to the left to overtake
- allow access into and from side roads, as blocking these will add to congestion.

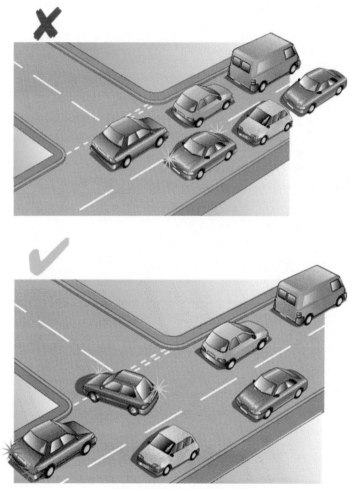

Do not block access to a side road

Driving in built-up areas

130. Narrow residential streets. You should drive slowly and carefully on streets where there are likely to be pedestrians, cyclists and parked cars. In some areas a 20 mph maximum speed limit may be in force.

Look out for
- vehicles emerging from junctions
- vehicles moving off
- car doors opening
- pedestrians
- children running out from between parked cars
- cyclists and motorcyclists.

131. Traffic calming measures. On some roads there are features such as road humps, chicanes and narrowings which are intended to slow you down. When you approach these features reduce your speed. Allow cyclists and motorcyclists room to pass through them. Maintain a reduced speed along the whole of the stretch of road within the calming measures. Give way to oncoming traffic if directed to do so by signs. You should not overtake other moving vehicles whilst in these areas.

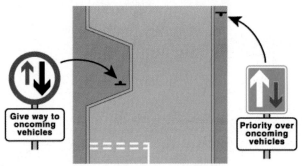

Chicanes may be used to slow traffic down

Country roads

132. Take extra care on country roads and reduce your speed at approaches to bends, which can be sharper than they appear, and at minor junctions and turnings, which may be partially hidden. Be prepared for pedestrians, horse riders and cyclists walking or riding in the road. You should also reduce your speed where country roads enter villages.

133. Single-track roads. These are only wide enough for one vehicle. They may have special passing places. If you see a vehicle coming towards you, or the driver behind wants to overtake, pull into a passing place on your left, or wait opposite a passing place on your right. Give way to vehicles coming uphill whenever you can. If necessary, reverse until you reach a passing place to let the other vehicle pass.

134. Do not park in passing places.

Using the road

General rules

135. Before moving off you should
- use all mirrors to check the road is clear
- look round to check the blind spots (the areas you are unable to see in the mirrors)
- signal if necessary before moving out
- look round for a final check.

Move off only when it is safe to do so.

Check the blind spot before moving off

136. Once moving you should
- keep to the left, unless road signs or markings indicate otherwise. The exceptions are when you want to overtake, turn right or pass parked vehicles or pedestrians in the road
- keep well to the left on right-hand bends. This will improve your view of the road and help avoid the risk of colliding with traffic approaching from the opposite direction
- keep both hands on the wheel, where possible. This will help you to remain in full control of the vehicle at all times
- be aware of other vehicles especially cycles and motorcycles. These are more difficult to see than larger vehicles and their riders are particularly vulnerable. Give them plenty of room, especially if you are driving a long vehicle or towing a trailer
- select a lower gear before you reach a long downhill slope. This will help to control your speed
- when towing, remember the extra length will affect overtaking and manoeuvring. The extra weight will also affect the braking and acceleration.

Mirrors

137. All mirrors should be used effectively throughout your journey. You should

- use your mirrors frequently so that you always know what is behind and to each side of you
- use them in good time before you signal or change direction or speed
- be aware that mirrors do not cover all areas and there will be blind spots. You will need to look round and check.

Remember: Mirrors – Signal – Manoeuvre

Overtaking

138. Before overtaking you should make sure

- the road is sufficiently clear ahead
- the vehicle behind is not beginning to overtake you
- there is a suitable gap in front of the vehicle you plan to overtake.

139. Overtake only when it is safe to do so. You should

- not get too close to the vehicle you intend to overtake
- use your mirrors, signal when it is safe to do so, take a quick sideways glance into the blind spot area and then start to move out
- not assume that you can simply follow a vehicle ahead which is overtaking; there may only be enough room for one vehicle
- move quickly past the vehicle you are overtaking, once you have started to overtake. Allow plenty of room. Move back to the left as soon as you can but do not cut in
- take extra care at night and in poor visibility when it is harder to judge speed and distance
- give way to oncoming vehicles before passing parked vehicles or other obstructions on your side of the road
- only overtake on the left if the vehicle in front is signalling to turn right, and there is room to do so
- stay in your lane if traffic is moving slowly in queues. If the queue on your right is moving more slowly than you are, you may pass on the left
- give motorcyclists, cyclists and horse riders at least as much room as you would a car when overtaking (see Rules 188, 189 and 191).

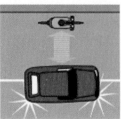

Remember: Mirrors – Signal – Manoeuvre

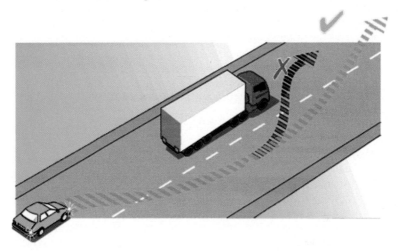

Do not cut in too quickly

140. Large vehicles. Overtaking these is more difficult.
You should

- drop back to increase your ability to see ahead. Getting too close to large vehicles will obscure your view of the road ahead and there may be another slow moving vehicle in front
- make sure that you have enough room to complete your overtaking manoeuvre before committing yourself. It takes longer to pass a large vehicle. If in doubt do not overtake
- not assume you can follow a vehicle ahead which is overtaking a long vehicle. If a problem develops, they may abort overtaking and pull back in.

141. You **MUST NOT** overtake

- if you would have to cross or straddle double white lines with a solid line nearest to you (but see Rule 108)
- if you would have to enter an area designed to divide traffic, if it is surrounded by a solid white line
- the nearest vehicle to a pedestrian crossing, especially when it has stopped to let pedestrians cross
- if you would have to enter a lane reserved for buses, trams or cycles during its hours of operation
- after a 'No Overtaking' sign and until you pass a sign cancelling the restriction.

Laws RTA 1988 sect 36, TSRGD regs 10,22,23 & 24, ZPPPCRGD reg 24

142. DO NOT overtake if there is any doubt, or where you cannot see far enough ahead to be sure it is safe. For example, when you are approaching

- a corner or bend
- a hump bridge
- the brow of a hill.

143. DO NOT overtake where you might come into conflict with other road users. For example

- approaching or at a road junction on either side of the road
- where the road narrows
- when approaching a school crossing patrol
- between the kerb and a bus or tram when it is at a stop
- where traffic is queuing at junctions or road works
- when you would force another vehicle to swerve or slow down
- at a level crossing
- when a vehicle is indicating right, even if you believe the signal should have been cancelled. Do not take a risk; wait for the signal to be cancelled.

144. Being overtaken. If a driver is trying to overtake you, maintain a steady course and speed, slowing down if necessary to let the vehicle pass. Never obstruct drivers who wish to pass. Speeding up or driving unpredictably while someone is overtaking you is dangerous. Drop back to maintain a two-second gap if someone overtakes and pulls into the gap in front of you.

145. Do not hold up a long queue of traffic, especially if you are driving a large or slow moving vehicle. Check your mirrors frequently, and if necessary, pull in where it is safe and let traffic pass.

Road junctions

146. Take extra care at junctions. You should

- watch out for cyclists, motorcyclists and pedestrians as they are not always easy to see
- watch out for pedestrians crossing a road into which you are turning. If they have started to cross they have priority, so give way
- watch out for long vehicles which may be turning at a junction ahead; they may have to use the whole width of the road to make the turn (see Rule 196)
- not assume, when waiting at a junction, that a vehicle coming from the right and signalling left will actually turn. Wait and make sure
- not cross or join a road until there is a gap large enough for you to do so safely.

147. You **MUST** stop behind the line at a junction with a 'Stop' sign and a solid white line across the road. Wait for a safe gap in the traffic before you move off.
Laws RTA 1988 sect 36 & TSRGD regs 10 & 16

148. The approach to a junction may have a 'Give Way' sign or a triangle marked on the road. You **MUST** give way to traffic on the main road when emerging from a junction with broken white lines across the road.
Laws RTA 1988 sect 36 & TSRGD regs 10(1), 16(1) & 25

149. Dual carriageways. When crossing or turning right, first assess whether the central reservation is deep enough to protect the full length of your vehicle.

- If it is, then you should treat each half of the carriageway as a separate road. Wait in the central reservation until there is a safe gap in the traffic on the second half of the road.
- If the central reservation is too shallow for the length of your vehicle, wait until you can cross both carriageways in one go.

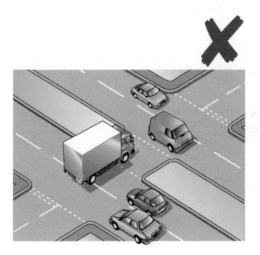

Assess your vehicle's length and do not obstruct traffic

150. Box junctions. These have criss-cross yellow lines painted on the road (see Road markings). You **MUST NOT** enter the box until your exit road or lane is clear. However, you may enter the box and wait when you want to turn right, and are only stopped from doing so by oncoming traffic, or by other vehicles waiting to turn right. At signalled roundabouts you **MUST NOT** enter the box unless you can cross over it completely without stopping.
Law TSRGD reg 10(1) & 29(2)

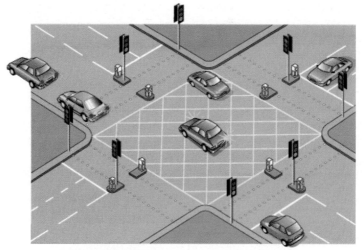

Enter a box junction only if your exit road is clear

Junctions controlled by traffic lights

151. You **MUST** stop behind the white 'Stop' line across your side of the road unless the light is green. If the amber light appears you may go on only if you have already crossed the stop line or are so close to it that to stop might cause an accident.
Laws RTA 1988 sect 36 & TSRGD regs 10 & 36

152. You **MUST NOT** move forward over the white line when the red light is showing. Only go forward when the traffic lights are green if there is room for you to clear the junction safely or you are taking up a position to turn right. If the traffic lights are not working, proceed with caution.
Laws RTA 1988 sect 36 & TSRGD regs 10 & 36

153. Green filter arrow. This indicates a filter lane only. Do not enter that lane unless you want to go in the direction of the arrow. You may proceed in the direction of the green arrow when it, or the full green light shows. Give other traffic, especially cyclists, time and room to move into the correct lane.

154. Advanced stop lines. Some junctions have advanced stop lines or bus advance areas to allow cycles and buses to be positioned ahead of other traffic. Motorists, including motorcyclists, **MUST** stop at the first white line reached, and should avoid encroaching on the marked area. If your vehicle has proceeded over the first white line at the time that the signal goes red, you **MUST** stop at the second white line, even if your vehicle is in the marked area. Allow cyclists and buses time and space to move off when the green signal shows.

Laws RTA 1988 sect 36 & TSRGD regs 10 & 43(2)

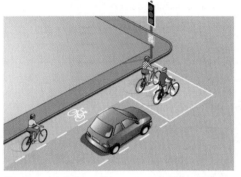

Do not encroach on the area marked for cyclists

Turning right

155. Well before you turn right you should

- use your mirrors to make sure you know the position and movement of traffic behind you
- give a right-turn signal
- take up a position just left of the middle of the road or in the space marked for traffic turning right
- leave room for other vehicles to pass on the left, if possible.

Position your vehicle correctly to avoid obstructing traffic

156. Wait until there is a safe gap between you and any oncoming vehicle. Watch out for cyclists, motorcyclists and pedestrians. Check your mirrors and blind spot again to make sure you are not being overtaken, then make the turn. Do not cut the corner. Take great care when turning into a main road; you will need to watch for traffic in both directions and wait for a safe gap.

Remember: Mirrors – Signal – Manoeuvre

157. When turning at a cross roads where an oncoming vehicle is also turning right, there is a choice of two methods

- turn right side to right side; keep the other vehicle on your right and turn behind it. This is generally the safest method as you have a clear view of any approaching traffic when completing your turn
- left side to left side, turning in front of each other. This can block your view of oncoming vehicles, so take extra care.

Road layout, markings or how the other vehicle is positioned can determine which course should be taken.

Turning right side to right side

Turning left side to left side

Turning left

158. Use your mirrors and give a left-turn signal well before you turn left. Do not overtake just before you turn left and watch out for traffic coming up on your left before you make the turn, especially if driving a large vehicle. Cyclists and motorcyclists in particular may be hidden from your view.

Do not cut in on cyclists

159. When turning
- keep as close to the left as is safe and practical
- give way to any vehicles using a bus lane, cycle lane or tramway from either direction.

Roundabouts

160. On approaching a roundabout take notice and act on all the information available to you, including traffic signs, traffic lights and lane markings which direct you into the correct lane. You should
- use **Mirrors – Signal – Manoeuvre** at all stages
- decide as early as possible which exit you need to take
- give an appropriate signal (see Rule 162). Time your signals so as not to confuse other road users
- get into the correct lane
- adjust your speed and position to fit in with traffic conditions
- be aware of the speed and position of all the traffic around you.

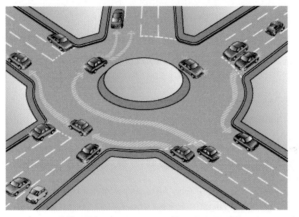

Follow the correct procedure at roundabouts

161. When reaching the roundabout you should
- give priority to traffic approaching from your right, unless directed otherwise by signs, road markings or traffic lights
- check whether road markings allow you to enter the roundabout without giving way. If so, proceed, but still look to the right before joining
- watch out for vehicles already on the roundabout; be aware they may not be signalling correctly or at all
- look forward before moving off to make sure traffic in front has moved off.

162. Signals and position, unless signs or markings indicate otherwise.

When taking the first exit
- signal left and approach in the left-hand lane
- keep to the left on the roundabout and continue signalling left to leave.

When taking any intermediate exit
- select the appropriate lane on approach to and on the roundabout, signalling where necessary
- stay in this lane until you need to alter course to exit the roundabout
- signal left after you have passed the exit before the one you want.

When taking the last exit or going full circle
- signal right and approach in the right-hand lane
- keep to the right on the roundabout until you need to change lanes to exit the roundabout
- signal left after you have passed the exit before the one you want.

When there are more than three lanes at the entrance to a roundabout, use the most appropriate lane on approach and through it.

163. In all cases watch out for and give plenty of room to
- pedestrians who may be crossing the approach and exit roads
- traffic crossing in front of you on the roundabout, especially vehicles intending to leave by the next exit
- traffic which may be straddling lanes or positioned incorrectly
- motorcyclists
- cyclists and horse riders who may stay in the left-hand lane and signal right if they intend to continue round the roundabout
- long vehicles (including those towing trailers) which might have to take a different course approaching or on the roundabout because of their length. Watch out for their signals.

164. Mini-roundabouts. Approach these in the same way as normal roundabouts. All vehicles **MUST** pass round the central markings except large vehicles which are physically incapable of doing so. Remember, there is less space to manoeuvre and less time to signal. Beware of vehicles making U-turns.
Laws RTA 1988 sect 36 & TSRGD regs 10(1) & 16(1)

165. At double mini-roundabouts treat each roundabout separately and give way to traffic from the right.

Treat each roundabout separately

166. Multiple roundabouts. At some complex junctions, there may be a series of mini-roundabouts at the intersections. Treat each mini-roundabout separately and follow the normal rules.

Pedestrian crossings

167. You **MUST NOT** park on a crossing or in the area covered by the zig-zag lines. You **MUST NOT** overtake the moving vehicle nearest the crossing or the vehicle nearest the crossing which has stopped to give way to pedestrians.

Laws ZPPPCRGD regs 18, 20 & 24, RTRA sect 25(5), TSRGD regs 10, 27 & 28

168. In queuing traffic, you should keep the crossing clear.

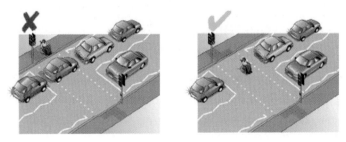

Keep the crossing clear

169. You should take extra care where the view of either side of the crossing is blocked by queuing traffic or incorrectly parked vehicles. Pedestrians may be crossing between stationary vehicles.

170. Allow pedestrians plenty of time to cross and do not harass them by revving your engine or edging forward.

171. Zebra crossings. As you approach a zebra crossing
- look out for people waiting to cross and be ready to slow down or stop to let them cross
- you **MUST** give way when someone has moved onto a crossing
- allow more time for stopping on wet or icy roads
- do not wave people across; this could be dangerous if another vehicle is approaching
- be aware of pedestrians approaching from the side of the crossing.

Law ZPPPCR reg 25

Signal-controlled crossings

172. Pelican crossings. These are signal-controlled crossings where flashing amber follows the red 'Stop' light. You **MUST** stop when the red light shows. When the amber light is flashing, you **MUST** give way to any pedestrians on the crossing. If the amber light is flashing and there are no pedestrians on the crossing, you may proceed with caution.

Laws ZPPPCRGD regs 23 & 26 & RTRA sect 25(5)

Allow pedestrians to cross when the amber light is flashing

173. Pelican crossings which go straight across the road are one crossing, even when there is a central island. You **MUST** wait for pedestrians who are crossing from the other side of the island.

Law ZPPPCRGD reg 26 & RTRA sect 25(5)

174. Give way to pedestrians who are still crossing after the signal for vehicles has changed to green.

175. Toucan and puffin crossings. These are similar to pelican crossings, but there is no flashing amber phase.

Reversing

176. Choose an appropriate place to manoeuvre. If you need to turn your car around, wait until you find a safe place. Try not to reverse or turn round in a busy road; find a quiet side road or drive round a block of side streets.

177. Do not reverse from a side road into a main road. When using a driveway, reverse in and drive out if you can.

178. Look carefully before you start reversing. You should
- use all your mirrors
- check the 'blind spot' behind you (the part of the road you cannot see easily in the mirrors)
- check there are no pedestrians, particularly children, cyclists, or obstructions in the road behind you
- look mainly through the rear window
- check all around just before you start to turn and be aware that the front of your vehicle will swing out as you turn
- get someone to guide you if you cannot see clearly.

Check all round when reversing

179. You **MUST NOT** reverse your vehicle further than necessary.
Law CUR reg 106

Road users requiring extra care

180. The most vulnerable road users are pedestrians, cyclists, motorcyclists and horse riders. It is particularly important to be aware of children, elderly and disabled people, and learner and inexperienced drivers and riders.

Pedestrians
181. In urban areas there is a risk of pedestrians, especially children, stepping unexpectedly into the road. You should drive with the safety of children in mind at a speed suitable for the conditions.

182. Drive carefully and slowly when
- in crowded shopping streets or residential areas
- driving past bus and tram stops; pedestrians may emerge suddenly into the road
- passing parked vehicles, especially ice cream vans; children are more interested in ice cream than traffic and may run into the road unexpectedly
- needing to cross a pavement; for example, to reach a driveway. Give way to pedestrians on the pavement
- reversing into a side road; look all around the vehicle and give way to any pedestrians who may be crossing the road
- turning at road junctions; give way to pedestrians who are already crossing the road into which you are turning
- the pavement is closed due to street repairs and pedestrians are directed to use the road.

Watch out for children in busy areas

183. Particularly vulnerable pedestrians. These include
- children and elderly pedestrians who may not be able to judge your speed and could step into the road in front of you. At 40 mph your vehicle will probably kill any pedestrians it hits. At 20 mph there is only a 1 in 20 chance of the pedestrian being killed. So kill your speed

- elderly pedestrians who may need more time to cross the road. Be patient and allow them to cross in their own time. Do not hurry them by revving your engine or edging forward
- blind and partially sighted people who may be carrying a white cane (white with a red band for deaf and blind people) or using a guide dog
- people with disabilities. Those with hearing problems may not be aware of your vehicle approaching. Those with walking difficulties require more time.

184. Near schools. Drive slowly and be particularly aware of young cyclists and pedestrians. In some places, there may be a flashing amber signal below the 'School' warning sign which tells you that there may be children crossing the road ahead. Drive very slowly until you are clear of the area.

185. Drive carefully when passing a stationary bus showing a 'School Bus' sign (see Vehicle markings) as children may be getting on or off.

186. You **MUST** stop when a school crossing patrol shows a 'Stop' for children sign (see Traffic signs).
Law RTRA sect 28

Motorcyclists and cyclists

187. It is often difficult to see motorcyclists and cyclists especially when they are coming up from behind, coming out of junctions and at roundabouts. Always look out for them when you are emerging from a junction.

Look out for motorcyclists at junctions

188. When passing motorcyclists and cyclists, give them plenty of room (see Rule 139). If they look over their shoulder whilst you are following them it could mean that they may soon attempt to turn right. Give them time and space to do so.

189. Motorcyclists and cyclists may suddenly need to avoid uneven road surfaces and obstacles such as draincovers or oily, wet or icy patches on the road. Give them plenty of room.

Other road users

190. Animals. When passing animals, drive slowly. Give them plenty of room and be ready to stop. Do not scare animals by sounding your horn or revving your engine. Look out for animals being led or ridden on the road and take extra care and keep your speed down at left-hand bends and on narrow country roads. If a road is blocked by a herd of animals, stop and switch off your engine until they have left the road. Watch out for animals on unfenced roads.

191. Horse riders. Be particularly careful of horses and riders, especially when overtaking. Always pass wide and slow. Horse riders are often children, so take extra care and remember riders may ride in double file when escorting a young or inexperienced horse rider. Look out for horse riders' signals and heed a request to slow down or stop. Treat all horses as a potential hazard and take great care.

192. Elderly drivers. Their reactions may be slower than other drivers. Make allowance for this.

193. Learners and inexperienced drivers. They may not be so skilful at reacting to events. Be particularly patient with learner drivers and young drivers. Drivers who have recently passed their test may display a 'new driver' plate or sticker.

Other vehicles

194. Emergency vehicles. You should look and listen for ambulances, fire engines, police or other emergency vehicles using flashing blue, red or green lights, headlights or sirens. When one approaches do not panic. Consider the route of the emergency vehicle and take appropriate action to let it pass. If necessary, pull to the side of the road and stop, but do not endanger other road users.

195. Powered vehicles used by disabled people. These small vehicles travel at a maximum speed of 8 mph. On a dual carriageway they **MUST** have a flashing amber light, but on other roads you may not have that advance warning.

Law RVLR reg 17(1)

196. Large vehicles. These may need extra road space to turn or to deal with a hazard that you are not able to see. If you are following a large vehicle, such as a bus or articulated lorry, be prepared to stop and wait if it needs room or time to turn.

Long vehicles need extra room

197. Large vehicles can block your view. Your ability to see and to plan ahead will be improved if you pull back to increase your separation distance.

198. Buses, coaches and trams. Give priority to these vehicles when you can do so safely, especially when they signal to pull away from stops. Look out for people getting off a bus or tram and crossing the road.

199. Electric vehicles. Be careful of electric vehicles such as milk floats and trams. Trams move quickly but silently and cannot steer to avoid you.

200. Vehicles with flashing amber lights. These warn of a slow-moving vehicle (such as a road gritter or recovery vehicle) or a vehicle which has broken down, so approach with caution.

Driving in adverse weather conditions

201. You **MUST** use headlights when visibility is seriously reduced, generally when you cannot see for more than 100 metres (328 feet). You may also use front or rear fog lights but you **MUST** switch them off when visibility improves (see Rule 211).
Law RVLR regs 25 & 27

Wet weather

202. In wet weather, stopping distances will be at least double those required for stopping on dry roads (see Rule 105 and Typical Stopping Distances diagram). This is because your tyres have less grip on the road. In wet weather

- you should keep well back from the vehicle in front. This will increase your ability to see and plan ahead
- if the steering becomes unresponsive, it probably means that water is preventing the tyres from gripping the road. Ease off the accelerator and slow down gradually
- the rain and spray from vehicles may make it difficult to see and be seen.

Icy and snowy weather

203. In winter check the local weather forecast for warnings of icy or snowy weather. **DO NOT** drive in these conditions unless your journey is essential. If it is, take great care. Carry a spade, warm clothing, a warm drink and emergency food in case your vehicle breaks down.

204. Before you set off

- you **MUST** be able to see, so clear all snow and ice from all your windows
- you **MUST** ensure that lights and number plates are clean
- make sure the mirrors are clear and the windows are de-misted thoroughly.

Laws CUR reg 30 & RVLR reg 23

Make sure your windscreen is completely clear

205. When driving in icy or snowy weather
- drive with care, even if the roads have been gritted
- keep well back from the vehicle in front as stopping distances can be ten times greater than on dry roads
- take care when overtaking gritting vehicles, particularly if you are riding a motorcycle
- watch out for snowploughs which may throw out snow on either side. Do not overtake them unless the lane you intend to use has been cleared
- be prepared for the road conditions changing over relatively short distances.

206. Drive extremely carefully when the roads are icy. Avoid sudden actions as these could cause a skid. You should
- drive at a slow speed in as high a gear as possible; accelerate and brake very gently
- drive particularly slowly on bends where skids are more likely. Brake progressively on the straight before you reach a bend. Having slowed down, steer smoothly round the bend, avoiding sudden actions
- check your grip on the road surface when there is snow or ice by choosing a safe place to brake gently. If the steering feels unresponsive this may indicate ice and your vehicle losing its grip on the road. When travelling on ice, tyres make virtually no noise.

Windy weather
207. High sided vehicles are most affected by windy weather, but strong gusts can also blow a car, cyclist or motorcyclist off course. This can happen at open stretches of road exposed to strong cross winds, or when passing bridges or gaps in hedges.

208. In very windy weather your vehicle may be affected by turbulence created by large vehicles. Motorcyclists are particularly affected, so keep well back from them when they are overtaking a high-sided vehicle.

Fog
209. Before entering fog check your mirrors then slow down. If the word 'Fog' is shown on a roadside signal but the road is clear, be prepared for a bank of fog or drifting patchy fog ahead. Even if it seems to be clearing, you can suddenly find yourself in thick fog.

210. When driving in fog you should
- use your lights as required in Rule 201
- keep a safe distance behind the vehicle in front. Rear lights can give a false sense of security

- be able to pull up within the distance you can see clearly. This is particularly important on motorways and dual carriageways, as vehicles are travelling faster
- use your windscreen wipers and demisters
- beware of other drivers not using headlights
- not accelerate to get away from a vehicle which is too close behind you
- check your mirrors before you slow down. Then use your brakes so that your brake lights warn drivers behind you that you are slowing down
- stop in the correct position at a junction with limited visibility and listen for traffic. When you are sure it is safe to emerge, do so positively and do not hesitate in a position that puts you directly in the path of approaching vehicles.

211. You **MUST NOT** use front or rear fog lights unless visibility is seriously reduced (see Rule 201) as they dazzle other road users and can obscure your brake lights. You **MUST** switch them off when visibility improves.
Law RVLR regs 25 & 27

Hot weather
212. Keep your vehicle well ventilated to avoid drowsiness. Be aware that the road surface may become soft or if it rains after a dry spell it may become slippery. These conditions could affect your steering and braking.

Waiting and parking

213. You **MUST NOT** wait or park where there are restrictions shown by
- yellow lines along the edge of the carriageway (see Road markings)
- school entrance markings on the carriageway.
The periods when restrictions apply are shown on upright signs, usually at intervals along the road, parallel to the kerb.
Law RTRA sects 5 & 8

Parking
214. Use off-street parking areas, or bays marked out with white lines on the road as parking places, wherever possible. If you have to stop on the road side

- stop as close as you can to the side
- do not stop too close to a vehicle displaying a Blue Badge, remember, they may need more room to get in or out
- you **MUST** switch off the engine, headlights and fog lights
- you **MUST** apply the handbrake before leaving the vehicle
- you **MUST** ensure you do not hit anyone when you open your door
- it is safer for your passengers (especially children) to get out of the vehicle on the side next to the kerb
- lock your vehicle.

Laws CSDPA sect 21, CUR reg 98,105 & 107, RVLR reg 27, RTA 1988 sect 42

Check before opening your door

215. You **MUST NOT** stop or park on

- the carriageway or the hard shoulder of a motorway except in an emergency (see Rule 244)
- a pedestrian crossing, including the area marked by the zig-zag lines (see Rule 167)
- a Clearway (see Traffic signs)
- a Bus Stop Clearway within its hours of operation
- taxi bays as indicated by upright signs and markings
- an Urban Clearway within its hours of operation, except to pick up or set down passengers (see Traffic signs)
- a road marked with double white lines, except to pick up or set down passengers
- a bus, tram or cycle lane during its period of operation
- a cycle track
- red lines, in the case of specially designated 'red routes', unless otherwise indicated by signs.

Laws MT(E&W)R regs 7 & 9, MT(S)R regs 6 & 8, ZPPPCRGD regs 18 & 20, RTRA sects 5 & 8, TSRGD regs 10, 26, 27 & 29(1), RTA 1988 sects 36 & 21(1)

216. You **MUST NOT** park in parking spaces reserved for specific users, such as Blue Badge holders or residents, unless entitled to do so.
Laws CSDPA sect 21 & RTRA sects 5 & 8

217. DO NOT park your vehicle or trailer on the road where it would endanger, inconvenience or obstruct pedestrians or other road users. For example, do not stop
- near a school entrance
- anywhere you would prevent access for Emergency Services
- at or near a bus stop or taxi rank
- on the approach to a level crossing
- opposite or within 10 metres (32 feet) of a junction, except in an authorised parking space
- near the brow of a hill or hump bridge
- opposite a traffic island or (if this would cause an obstruction) another parked vehicle
- where you would force other traffic to enter a tram lane
- where the kerb has been lowered to help wheelchair users
- in front of an entrance to a property
- on a bend.

218. DO NOT park partially or wholly on the pavement unless signs permit it. Parking on the pavement can obstruct and seriously inconvenience pedestrians, people in wheelchairs, the visually impaired and people with prams or pushchairs.

219. Controlled Parking Zones. The zone entry signs indicate the times when the waiting restrictions within the zone are in force. Parking may be allowed in some places at other times. Otherwise parking will be within separately signed and marked bays.

220. Goods vehicles. Vehicles with a maximum laden weight of over 7.5 tonnes (including any trailer) **MUST NOT** be parked on a verge, pavement or any land situated between carriageways, without police permission. The only exception is when parking is essential for loading and unloading, in which case the vehicle **MUST NOT** be left unattended.
Law RTA 1988 sect 19

221. Loading and unloading. Do not load or unload where there are yellow markings on the kerb and upright signs advise restrictions are in place (see Road markings). This may be permitted where parking is otherwise restricted. On red routes, specially marked and signed bays indicate where and when loading and unloading is permitted.
Law RTRA sects 5 & 8

Parking at night

222. You **MUST NOT** park on a road at night facing against the direction of the traffic flow unless in a recognised parking space.

Laws CUR reg 101 & RVLR reg 24

223. All vehicles **MUST** display parking lights when parked on a road or a lay-by on a road with a speed limit greater than 30 mph.

Law RVLR reg 24

224. Cars, goods vehicles not exceeding 1525kg unladen, invalid carriages and motorcycles may be parked without lights on a road (or lay-by) with a speed limit of 30 mph or less if they are

- at least 10 metres (32 feet) away from any junction, close to the kerb and facing in the direction of the traffic flow
- in a recognised parking place or lay-by.

Other vehicles and trailers, and all vehicles with projecting loads, **MUST NOT** be left on a road at night without lights.

Law RVLR reg 24

225. Parking in fog. It is especially dangerous to park on the road in fog. If it is unavoidable, leave your parking lights or sidelights on.

226. Parking on hills. If you park on a hill you should

- park close to the kerb and apply the handbrake firmly
- select a forward gear and turn your steering wheel away from the kerb when facing uphill
- select reverse gear and turn your steering wheel towards the kerb when facing downhill
- use 'park' if your car has an automatic gearbox.

Motorways

Many other Rules apply to motorway driving, either wholly or in part: Rules 43, 67–105, 109–113, 118, 122, 126–128, 135, 137, 194, 196, 200, 201–212, 248–252, 254–264.

General

227. Prohibited vehicles. Motorways **MUST NOT** be used by pedestrians, holders of provisional car or motorcycle driving licences unless exempt, riders of motorcycles under 50cc, cyclists and horse riders. Certain slow-moving vehicles and those carrying oversized loads (except by special permission), agricultural vehicles and most invalid carriages are also prohibited.

Laws HA 1980 sects 16, 17 & sch 4, MT(E&W)R reg 4, MT(E&W)(A)R, R(S)A sects 7 ,8 & sch 3 & MT(S)R reg 10

228. Traffic on motorways usually travels faster than on other roads, so you have less time to react. It is especially important to use your mirrors earlier and look much further ahead than you would on other roads.

Motorway signals

229. Motorway signals (see Light signals controlling traffic) are used to warn you of a danger ahead. For example, there may be an accident, fog, or a spillage, which you may not immediately be able to see.

230. Signals situated on the central reservation apply to all lanes. On very busy stretches, signals may be overhead with a separate signal for each lane.

231. Amber flashing lights. These warn of a hazard ahead. The signal may show a temporary maximum speed limit, lanes that are closed or a message such as 'Fog'. Adjust your speed and look out for the danger until you pass a signal which is not flashing or one that gives the 'All clear' sign and you are sure it is safe to increase your speed.

232. Red flashing lights. If red lights on the overhead signals flash above your lane (there may also be a red 'X') you **MUST NOT** go beyond the signal in that lane. If red lights flash on a signal in the central reservation or at the side of the road, you **MUST NOT** go beyond the signal in any lane.

Laws RTA 1988 sect 36 & TSRGD regs 10 & 38

Driving on the motorway
Joining the motorway
233. When you join the motorway you will normally approach it from a road on the left (a slip road) or from an adjoining motorway. You should

- give priority to traffic already on the motorway
- check the traffic on the motorway and adjust your speed to fit safely into the traffic flow in the left-hand lane
- not cross solid white lines that separate lanes
- stay on the slip road if it continues as an extra lane on the motorway
- remain in the left-hand lane long enough to adjust to the speed of traffic before considering overtaking.

On the motorway
234. When you can see well ahead and the road conditions are good, you should

- drive at a steady cruising speed which you and your vehicle can handle safely and is within the speed limit (see Rule 103 and Speed Limits diagram)
- keep a safe distance from the vehicle in front and increase the gap on wet or icy roads, or in fog (see Rules 105 & 210).

235. You **MUST NOT** exceed 70 mph, or the maximum speed limit permitted for your vehicle (see Rule 103 and Speed Limits diagram). If a lower speed limit is in force, either permanently or temporarily, at roadworks for example, you **MUST NOT** exceed the lower limit. On some motorways, mandatory motorway signals (which display the speed within a red ring) are used to vary the maximum speed limit to improve traffic flow. You **MUST NOT** exceed this speed limit.
Law RTRA sects 17, 86, 89 & sch 6

236. The monotony of driving on a motorway can make you feel sleepy. To minimise the risk, follow the advice in Rule 80.

237. You **MUST NOT** reverse, cross the central reservation, or drive against the traffic flow. If you have missed your exit, or have taken the wrong route, carry on to the next exit.
Laws MT(E&W)R regs 6, 7 & 10 & MT(S)R regs 4, 5, 7 & 9

Lane discipline
238. You should drive in the left-hand lane if the road ahead is clear. If you are overtaking a number of slower moving vehicles it may be safer to remain in the centre or outer lanes until the manoeuvre is completed rather than continually changing lanes. Return to the left-hand lane once you have overtaken all the vehicles or if you are delaying traffic

behind you. Slow moving or speed restricted vehicles should always remain in the left-hand lane of the carriageway unless overtaking. You **MUST NOT** drive on the hard shoulder except in an emergency or if directed to do so by signs.

Laws MT(E&W)R regs 5, 9 & 16(1)(a) & MT(S)R regs 4, 8 & 14(1)(a)

239. The right-hand lane of a motorway with three or more lanes **MUST NOT** be used (except in prescribed circumstances) if you are driving
- any vehicle drawing a trailer
- a goods vehicle with a maximum laden weight over 7.5 tonnes
- a passenger vehicle with a maximum laden weight exceeding 7.5 tonnes constructed or adapted to carry more than eight seated passengers in addition to the driver.

Laws MT(E&W)R reg 12 & MT(S)R reg 11

240. Approaching a junction. Look well ahead for signals or signs. Direction signs may be placed over the road. If you need to change lanes, do so in good time. At some junctions a lane may lead directly off the motorway. Only get in that lane if you wish to go in the direction indicated on the overhead signs.

Overtaking

241. Do not overtake unless you are sure it is safe to do so. Overtake only on the right. You should
- check your mirrors
- take time to judge the speeds correctly
- make sure that the lane you will be joining is sufficiently clear ahead and behind
- take a quick sideways glance into the blind spot area to verify the position of a vehicle that may have disappeared from your view in the mirror
- remember that traffic may be coming up behind you very quickly. Check your mirrors carefully. When it is safe to do so, signal in plenty of time, then move out
- ensure you do not cut in on the vehicle you have overtaken
- be especially careful at night and in poor visibility when it is harder to judge speed and distance.

242. Do not overtake on the left or move to a lane on your left to overtake. In congested conditions, where adjacent lanes of traffic are moving at similar speeds, traffic in left-hand lanes may sometimes be moving faster than traffic to the right. In these conditions you may keep up with the traffic in your lane even if this means passing traffic in the lane to your right. Do not weave in and out of lanes to overtake.

243. You **MUST NOT** use the hard shoulder for overtaking.
Laws MT(E&W)R regs 5 & 9 & MT(S)R regs 4 & 8

Stopping
244. You **MUST NOT** stop on the carriageway, hard shoulder, slip road, central reservation or verge except in an emergency, or when told to do so by the police, an emergency sign or by flashing red light signals.
Laws MT(E&W)R regs 7(1), 9, 10 & 16 & MT(S)R regs 6(1), 8, 9 & 14

245. You **MUST NOT** pick up or set down anyone, or walk on a motorway, except in an emergency.
Laws RTRA sect 17 & MT(E&W)R reg 15

Leaving the motorway
246. Unless signs indicate that a lane leads directly off the motorway, you will normally leave the motorway by a slip road on your left. You should
- watch for the signs letting you know you are getting near your exit
- move into the left-hand lane well before reaching your exit
- signal left in good time and reduce your speed on the slip road as necessary.

247. On leaving the motorway or using a link road between motorways, your speed may be higher than you realise – 50 mph may feel like 30 mph. Check your speedometer and adjust your speed accordingly. Some slip roads and link roads have sharp bends, so you will need to slow down.

Breakdowns and accidents

Breakdowns

248. If your vehicle breaks down, think first of other road users and

- get your vehicle off the road if possible
- warn other traffic by using your hazard warning lights if your vehicle is causing an obstruction
- put a warning triangle on the road at least 45 metres (147 feet) behind your broken down vehicle on the same side of the road, or use other permitted warning devices if you have them. Always take great care when placing them, but never use them on motorways
- keep your sidelights on if it is dark or visibility is poor
- do not stand (or let anybody else stand), between your vehicle and oncoming traffic
- at night or in poor visibility do not stand where you will prevent other road users seeing your lights.

Additional rules for the motorway

249. If your vehicle develops a problem, leave the motorway at the next exit or pull into a service area. If you cannot do so, you should

- pull on to the hard shoulder and stop as far to the left as possible, with your wheels turned to the left
- try to stop near an emergency telephone (situated at approximately one mile intervals along the hard shoulder)
- leave the vehicle by the left-hand door and ensure your passengers do the same. You **MUST** leave any animals in the vehicle or, in an emergency, keep them under proper control on the verge
- do not attempt even simple repairs
- ensure that passengers keep away from the carriageway and hard shoulder, and that children are kept under control
- walk to an emergency telephone on your side of the carriageway (follow the arrows on the posts at the back of the hard shoulder)– the telephone is free of charge and connects directly to the police. Use these in preference to a mobile phone (see Rule 257)
- give full details to the police; also inform them if you are a vulnerable motorist such as a woman travelling alone
- return and wait near your vehicle (well away from the carriageway and hard shoulder)
- if you feel at risk from another person, return to your vehicle by a left-hand door and lock all doors. Leave your vehicle again as soon as you feel this danger has passed.

Laws MT(E&W)R reg 14 & MT(S)R reg 12

Keep well back from the hard shoulder

250. Before you rejoin the carriageway after a breakdown, build up speed on the hard shoulder and watch for a safe gap in the traffic. Be aware that other vehicles may be stationary on the hard shoulder.

251. If you cannot get your vehicle on to the hard shoulder
- do not attempt to place any warning device on the carriageway
- switch on your hazard warning lights
- leave your vehicle only when you can safely get clear of the carriageway.

Disabled drivers
252. If you have a disability which prevents you from following the above advice you should
- stay in your vehicle
- switch on your hazard warning lights
- display a 'Help' pennant or, if you have a car or mobile telephone, contact the emergency services and be prepared to advise them of your location.

Obstructions
253. If anything falls from your vehicle (or any other vehicle) on to the road, stop and retrieve it only if it is safe to do so.

254. Motorways. On a motorway do not try to remove the obstruction yourself. Stop at the next emergency telephone and call the police.

The Highway Code

Accidents

255. Warning signs or flashing lights. If you see or hear emergency vehicles in the distance be aware there may be an accident ahead.

256. When passing the scene of an accident do not be distracted or slow down unnecessarily (for example if an accident is on the other side of a dual carriageway). This may cause another accident or traffic congestion, but see Rule 257.

257. If you are involved in an accident or stop to give assistance
- use your hazard warning lights to warn other traffic
- ask drivers to switch off their engines and stop smoking
- arrange for the emergency services to be called immediately with full details of the accident location and any casualties (on a motorway, use the emergency telephone which allows easy location by the emergency services. If you use a mobile phone, first make sure you have identified your location from the marker posts on the side of the hard shoulder)
- move uninjured people away from the vehicles to safety; on a motorway this should, if possible, be well away from the traffic, the hard shoulder and the central reservation
- do not move injured people from their vehicles unless they are in immediate danger from fire or explosion
- do not remove a motorcyclist's helmet unless it is essential to do so
- be prepared to give first aid as shown in Annexe 7: First aid on the road
- stay at the scene until emergency services arrive.

If you are involved in any other medical emergency on the motorway you should contact the emergency services in the same way.

Accidents involving dangerous goods

258. Vehicles carrying dangerous goods in packages will be marked with plain orange reflective plates. Road tankers and vehicles carrying tank containers of dangerous goods will have hazard warning plates (see Vehicle markings).

259. If an accident involves a vehicle containing dangerous goods, follow the advice in Rule 257 and, in particular

- switch off engines and **DO NOT SMOKE**
- keep well away from the vehicle and do not be tempted to try to rescue casualties as you yourself could become one
- call the emergency services and give as much information as possible about the labels and markings on the vehicle.
  DO NOT use a mobile phone close to a vehicle carrying flammable loads.

Documentation

260. If you are involved in an accident which causes damage or injury to any other person, vehicle, animal or property, you **MUST**

- stop
- give your own and the vehicle owner's name and address, and the registration number of the vehicle, to anyone having reasonable grounds for requiring them
- if you do not give your name and address at the time of the accident, report the accident to the police as soon as reasonably practicable, and in any case within 24 hours.

Law RTA 1988 sect 170

261. If another person is injured and you do not produce your insurance certificate at the time of the accident to a police officer or to anyone having reasonable grounds to request it, you **MUST**

- report the accident to the police as soon as possible and in any case within 24 hours
- produce your insurance certificate for the police within seven days.

Law RTA 1988 sect 170

Road works

262. When the 'Road Works Ahead' sign is displayed, you will need to be more watchful and look for additional signs providing more specific instructions.

- You **MUST NOT** exceed any temporary maximum speed limit.
- Use your mirrors and get into the correct lane for your vehicle in good time and as signs direct.
- Do not switch lanes to overtake queuing traffic.
- Do not drive through an area marked off by traffic cones.
- Watch out for traffic entering or leaving the works area, but do not be distracted by what is going on there.
- Bear in mind that the road ahead may be obstructed by the works or by slow moving or stationary traffic.

Law RTRA sect 16

Additional rules for high speed roads

263. Take special care on motorways and other high speed dual carriageways.

- One or more lanes may be closed to traffic and a lower speed limit may apply.
- Works vehicles that are slow moving or stationary with a large 'Keep Left' or 'Keep Right' sign on the back are sometimes used to close lanes for repairs.
- Check mirrors, slow down and change lanes if necessary.
- Keep a safe distance from the vehicle in front (see Rule 105).

264. Contraflow systems mean that you may be travelling in a narrower lane than normal and with no permanent barrier between you and oncoming traffic. The hard shoulder may be used for traffic, but be aware that there may be broken down vehicles ahead of you. Keep a good distance from the vehicle ahead and observe any temporary speed limits.

Railway level crossings

265. A level crossing is where a road crosses a railway line. Approach and cross it with care. Never drive on to a crossing until the road is clear on the other side and do not get too close to the car in front. Never stop or park on, or near, a crossing.

Controlled crossings

266. Most crossings have traffic light signals with a steady amber light, twin flashing red stop lights (see Light signals controlling traffic and Traffic signs) and an audible alarm for pedestrians. They may have full, half or no barriers.

- You **MUST** always obey the flashing red stop lights.
- You **MUST** stop behind the white line across the road.
- Keep going if you have already crossed the white line when the amber light comes on.
- You **MUST** wait if a train goes by and the red lights continue to flash. This means another train will be passing soon.
- Only cross when the lights go off and barriers open.
- Never zig-zag around half-barriers, they lower automatically because a train is approaching.
- At crossings where there are no barriers, a train is approaching when the lights show.

Laws RTA 1988 sect 36 & TSRGD regs 10 & 40

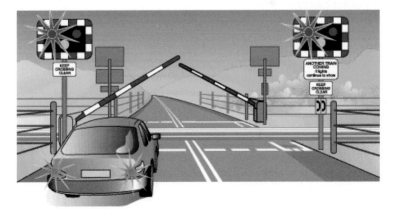

Stop when the traffic lights show

267. Railway telephones. If you are driving a large or slow-moving vehicle, or herding animals, a train could arrive before you are clear of the crossing. You **MUST** obey any sign instructing you to use the railway telephone to obtain permission to cross. You **MUST** also telephone when clear of the crossing.
Laws RTA 1988 sect 36 & TSRGD regs 10 & 16(1)

268. Crossings without traffic lights. Vehicles should stop and wait at the barrier or gate when it begins to close and not cross until the barrier or gate opens.

User-operated gates or barriers
269. Some crossings have 'Stop' signs and small red and green lights. You **MUST NOT** cross when the red light is showing, only cross if the green light is on. If crossing with a vehicle, you should
- open the gates or barriers on both sides of the crossing
- check that the green light is still on and cross quickly
- close the gates or barriers when you are clear of the crossing.
Laws RTA 1988 sect 36 & TSRGD regs 10 & 52(2)

270. If there are no lights, follow the procedure in Rule 269. Stop, look both ways and listen before you cross. If there is a railway telephone, always use it to contact the signal operator to make sure it is safe to cross. Inform the signal operator again when you are clear of the crossing.

Open crossings
271. These have no gates, barriers, attendant or traffic lights but will have a 'Give Way' sign. You should look both ways, listen and make sure there is no train coming before you cross.

Accidents and breakdowns
272. If your vehicle breaks down, or if you have an accident on a crossing you should
- get everyone out of the vehicle and clear of the crossing immediately
- use a railway telephone if available to tell the signal operator. Follow the instructions you are given
- move the vehicle clear of the crossing if there is time before a train arrives. If the alarm sounds, or the amber light comes on, leave the vehicle and get clear of the crossing immediately.

Tramways

273. You **MUST NOT** enter a road, lane or other route reserved for trams. Take extra care where trams run along the road. The width taken up by trams is often shown by tram lanes marked by white lines, yellow dots or by a different type of road surface. Diamond-shaped signs give instructions to tram drivers only.
Law RTRA sects 5 & 8

274. Take extra care where the track crosses from one side of the road to the other and where the road narrows and the tracks come close to the kerb. Tram drivers usually have their own traffic signals and may be permitted to move when you are not. Always give way to trams. Do not try to race or overtake them.

275. You **MUST NOT** park your vehicle where it would get in the way of trams or where it would force other drivers to do so.
Law RTRA sects 5 & 8

276. Tram stops. Where the tram stops at a platform, either in the middle or at the side of the road, you **MUST** follow the route shown by the road signs and markings. At stops without platforms you **MUST NOT** drive between a tram and the left-hand kerb when a tram has stopped to pick up passengers.
Law RTRA sects 5 & 8

277. Look out for pedestrians, especially children, running to catch a tram approaching a stop.

278. Cyclists and motorcyclists should take extra care when riding close to or crossing the tracks, especially if the rails are wet. It is safest to cross the tracks directly at right angles.

Light signals controlling traffic

Traffic Light Signals

RED means 'Stop'. Wait behind the stop line on the carriageway

RED AND AMBER also means 'Stop'. Do not pass through or start until GREEN shows

GREEN means you may go on if the way is clear. Take special care if you intend to turn left or right and give way to pedestrians who are crossing

AMBER means 'Stop' at the stop line. You may go on only if the AMBER appears after you have crossed the stop line or are so close to it that to pull up might cause an accident

A GREEN ARROW may be provided in addition to the full green signal if movement in a certain direction is allowed before or after the full green phase. If the way is clear you may go but only in the direction shown by the arrow. You may do this whatever other lights may be showing. White light signals may be provided for trams

Flashing red lights

Alternately flashing red lights mean YOU MUST STOP

At level crossings, lifting bridges, airfields, fire stations, etc

Motorway signals

Do not proceed further in this lane

Change lane

Reduced visibilty ahead

Lane ahead closed

Temporary maximum speed limit and information message

Leave motorway at next exit

Temporary maximum speed limit

End of restriction

Lane control signals

Green arrow – lane available to traffic facing the sign.
Red crosses – lane closed to traffic facing the sign.
White diagonal arrow – change lanes in direction shown.

Signals to other road users

Direction indicator signals

I intend to move out to the
right or turn right

I intend to move in to the left or
turn left or stop on the left

Brake light signals

I am applying the brakes

Reversing light signals

I intend to reverse

These signals should not be used except for the purposes described.

Arm signals

For use when direction indicator signals are not used, or when necessary to reinforce direction indicator signals and stop lights. *Also for use by pedal cyclists and those in charge of horses.*

I intend to move in to
the left or turn left

I intend to move out to
the right or turn right

I intend to slow
down or stop

Signals by authorised persons

Stop

Traffic approaching from the front

Traffic approaching from both front and behind

Traffic approaching from behind

To beckon traffic on

From the side

From the front

From behind*

Arm signals to persons controlling traffic

I want to go straight on

I want to turn left; use either hand

I want to turn right

*In Wales, bilingual signs appear on emergency services vehicles and clothing

Traffic signs

Signs giving orders

Signs with red circles are mostly prohibitive. Plates below signs qualify their message.

Entry to 20 mph zone

End of 20 mph zone

School crossing patrol

Maximum speed

National speed limit applies

Stop and give way

Give way to traffic on major road

No vehicles except bicycles being pushed

Give priority to vehicles from opposite direction

No vehicle or combination of vehicles over length shown

No vehicles over height shown

No vehicles over width shown

No goods vehicles over maximum gross weight shown (in tonnes) except for loading and unloading

No overtaking

No motor vehicles

Manually operated temporary STOP and GO signs

No buses (over 8 passenger seats)

No towed caravans

No vehicles carrying explosives

No right turn

No left turn

No U-turns

No cycling

No entry for vehicular traffic

No waiting

No stopping (Clearway)

No stopping during times shown except for as long as necessary to set down or pick up passengers

No vehicles over maximum gross weight shown (in tonnes)

Parking restricted to permit holders

No stopping during period indicated except for buses

Note: Although *The Highway Code* shows many of the signs commonly in use, a comprehensive explanation of the signing system is given in the Department's booklet *Know Your Traffic Signs*, which is on sale at booksellers. The booklet also illustrates and explains the vast majority of signs the road user is likely to encounter. The signs illustrated in *The Highway Code* are not all drawn to the same scale. In Wales, bilingual versions of some signs are used including Welsh and English versions of place names. Some older designs of signs may still be seen on the roads.

Signs with blue circles but no red border mostly give positive instruction.

One-way traffic (note: compare circular 'Ahead only' sign) | Ahead only | Turn left ahead (right if symbol reversed) | Turn left (right if symbol reversed) | Keep left (right if symbol reversed) | Route to be used by pedal cycles only

Segregated pedal cycle and pedestrian route | Minimum speed | End of minimum speed | Mini-roundabout (roundabout circulation – give way to vehicles from the immediate right) | Vehicles may pass either side to reach same destination

Buses and cycles only | Trams only | Pedestrian crossing point over tramway | With-flow bus and cycle lane | Contra-flow bus lane

With-flow pedal cycle lane

Warning signs **Mostly triangular**

Distance to 'STOP' line ahead | Crossroads | Junction on bend ahead | T-junction | Staggered junction | Distance to 'Give Way' line ahead

The priority through route is indicated by the broader line.

Sharp deviation of route to left (or right if chevrons reversed) | Double bend first to left (symbol may be reversed) | Bend to right (or left if symbol reversed) | Roundabout | Uneven road | Plate below some signs

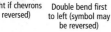 

Dual carriage-way ends | Road narrows on right (left if symbol reversed) | Road narrows on both sides | Two-way traffic crosses one-way road | Two-way traffic straight ahead

Traffic signals | Traffic signals not in use | Slippery road | Steep hill downwards | Steep hill upwards

Gradients may be shown as a ratio i.e. 20% = 1:5

Warning signs – continued

School crossing patrol ahead (some signs have amber lights which flash when children are crossing)

Frail (or blind or disabled if shown) pedestrians likely to cross road ahead

Pedestrians in road ahead

Pedestrian crossing

Traffic queues likely ahead

Cycle route ahead

Side winds

Hump bridge

Worded warning sign

Risk of ice

Risk of grounding

Light signals ahead at level crossing, airfield or bridge

Level crossing with barrier or gate ahead

Level crossing without barrier or gate ahead

Level crossing without barrier

Trams crossing ahead

Cattle

Wild animals

Wild horses or ponies

Accompanied horses or ponies

Quayside or river bank

Opening or swing bridge ahead

Low-flying aircraft or sudden aircraft noise

Falling or fallen rocks

Available width of headroom indicated

Overhead electric cable; plate indicates maximum height of vehicles which can pass safely

Tunnel ahead

Distance over which road humps extend

Other danger; plate indicates nature of danger

Soft verges

Direction signs Mostly rectangular

Signs on motorways – blue backgrounds

At a junction leading directly into a motorway (junction number may be shown on a black background)

On approaches to junctions (junction number on black background)

Route confirmatory sign after junction

Downward pointing arrows mean 'Get in lane'
The left-hand lane leads to a different destination from the other lanes

The panel with the inclined arrow indicates the destinations which can be reached by leaving the motorway at the next junction

Signs on primary routes – green backgrounds

On approaches to junctions

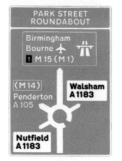

On approaches to junctions

Blue panels indicate that the motorway starts at the junction ahead.
Motorways shown in brackets can also be reached along the route indicated.
White panels indicate local or non–primary routes leading from the junction ahead.
Brown panels show the route to tourist attractions.
The name of the junction may be shown at the top of the sign.
The aircraft symbol indicates the route to an airport.
A symbol may be included to warn of a hazard or restriction along that route.

Route confirmatory sign after junction

At the junction

On approach to a junction in Wales (bilingual)

Signs on non-primary and local routes – black borders

On approaches to junctions

Green panels indicate that the primary route starts at the junction ahead. Route numbers on a blue background show the direction to a motorway. Route numbers on a green background show the direction to a primary route.

At the junction

Direction to toilets with access for the disabled

Other direction signs

Picnic site

Wrest Park
Ancient monument in the care of English Heritage

Direction to camping and caravan site

Advisory route for lorries

Zoo
Tourist attraction

Route for pedal cycles forming part of a network

Public library Council offices
Route for pedestrians

Northtown
Diversion route

Marton 3
Recommended route for pedal cycles to place shown

Symbols showing emergency diversion route for motorway and other main road traffic

HR
Holiday route

Saturday only
Direction to a car park

Information signs All rectangular

M 62
Start of motorway and point from which motorway regulations apply

Area in which cameras are used to enforce traffic regulations

Priority over oncoming vehicles
Traffic has priority over oncoming vehicles

No through road for vehicles

H A & E not 24 hrs
Hospital ahead with Accident and Emergency facilities

i Tourist information
Tourist information point

End of motorway

P
Parking place for solo motorcycles

Low bridge 2 miles ahead 4.4 m 14'6"
Advance warning of restriction or prohibition ahead

'Countdown' markers at exit from motorway (each bar represents 100 yards to the exit). Green-backed markers may be used on primary routes and white-backed markers with black bars on other routes. At approaches to concealed level crossings white-backed markers with red bars may be used. Although these will be erected at equal distances the bars do not represent 100 yard intervals.

GOOD FOOD Puddleworth services ½ m Petrol 85p
Motorway service area sign showing the operator's name

Recommended route for pedal cycles

local taxi
With-flow bus lane ahead which pedal cycles and taxis may also use

Bus lane
Bus lane on road at junction ahead

Appropriate traffic lanes at junction ahead

Controlled ZONE Mon - Fri 8.30 am - 6.30 pm Saturday 8.30 am - 1.30 pm
Entrance to controlled parking zone

Zone ENDS
End of controlled parking zone

Road works signs

Road works

Loose chippings

Road works 1 mile ahead

End of road works and any temporary restrictions

Temporary hazard at road works

Temporary lane closure (the number and position of arrows and red bars may be varied according to lanes open and closed)

Lane restrictions at road works ahead

One lane crossover at contraflow road works

Signs used on the back of slow-moving or stationary vehicles warning of a lane closed ahead by a works vehicle. There are no cones on the road

Slow-moving or stationary works vehicle blocking a traffic lane. Pass in the direction shown by the arrow

Mandatory speed limit ahead

Road markings Across the carriageway

Stop line at signals or police control

Stop line at 'Stop' sign

Stop line for pedestrians at a level crossing

Give way to traffic on major road

Give way to traffic from the right at a roundabout

Give way to traffic from the right at a mini-roundabout

Along the carriageway

Edge line

Centre line
See Rule 106

Hazard warning line
See Rule 106

Double white lines
See Rules 107 and 108

Diagonal hatching
See Rule 109

Lane line
See Rule 110

Along the edge of the carriageway

Waiting restrictions

Waiting restrictions indicated by yellow lines apply to the carriageway, pavement and verge. You may stop to load or unload (unless there are also loading restrictions as described below) or while passengers board or alight. Double yellow lines mean no waiting at any time, unless there are signs that specifically indicate seasonal restrictions. The times at which the restrictions apply for other road markings are shown on nearby plates or on entry signs to controlled parking zones. If no days are shown on the signs, the restrictions are in force every day including Sundays and Bank Holidays. White bay markings and upright signs (see below) indicate where parking is allowed.

Red Route stopping controls

Red lines are used on some roads instead of yellow lines. In London the double and single red lines used on Red Routes indicate that stopping to park, load/unload or to board and alight from a vehicle (except for a licensed taxi or if you hold a Blue Badge) is prohibited. The red lines apply to the carriageway, pavement and verge. The times that the red line prohibitions apply are shown on nearby signs, but the double red line ALWAYS means no stopping at any time. On Red Routes you may stop to park, load/unload in specially marked boxes and adjacent signs specify the times and purposes and duration allowed. A box MARKED IN RED indicates that it may only be available for the purpose specified for part of the day (eg between busy peak periods). A box MARKED IN WHITE means that it is available throughout the day.

RED AND SINGLE YELLOW LINES CAN ONLY GIVE A GUIDE TO THE RESTRICTIONS AND CONTROLS IN FORCE AND SIGNS, NEARBY OR AT A ZONE ENTRY, MUST BE CONSULTED.

No waiting at any time

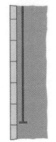

No waiting during times shown on sign

Waiting is limited to the times and duration shown

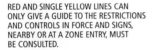

No stopping at any time

No stopping during times shown on sign

Parking is limited at the times for duration shown

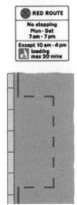

Only loading may take place at the times shown for up to a maximum duration of 20 minutes

On the kerb or at the edge of the carriageway

Loading restrictions on roads other than Red Routes

Yellow marks on the kerb or at the edge of the carriageway indicate that loading or unloading is prohibited at the times shown on the nearby black and white plates. You may stop while passengers board or alight. If no days are indicated on the signs the restrictions are in force every day including Sundays and Bank Holidays. ALWAYS CHECK THE TIMES SHOWN ON THE PLATES.

Lengths of road reserved for vehicles loading and unloading are indicated by a white 'bay' marking with the words 'Loading Only' and a sign with the white on blue 'trolley' symbol. This sign also shows whether loading and unloading is restricted to goods vehicles and the times at which the bay can be used. If no times or days are shown it may be used at any time. Vehicles may not park here if they are not loading or unloading.

No loading at any time

No loading or unloading at any time

No loading Mon - Sat 8.30 am - 6.30 pm

No loading or unloading at the times shown

Loading only

Loading bay

Other road markings

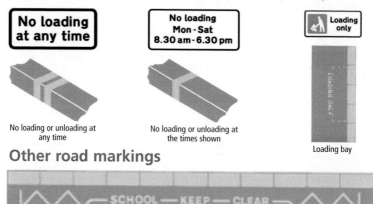

Keep entrance clear of stationary vehicles, even if picking up or setting down children

Warning of 'Give Way' just ahead

Parking space reserved for vehicles named

See Rule 215

See rule 120

Box junction See Rule 150

Do not block that part of the carriageway indicated

Indication of traffic lanes

Vehicle markings

Large goods vehicle rear markings

Motor vehicles over 7500 kilograms maximum gross weight and trailers over 3500 kilograms maximum gross weight

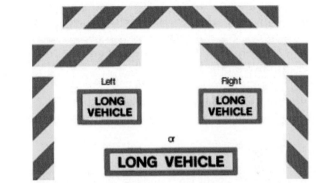

School bus (displayed in front or rear window of bus or coach)

Left

Right

LONG VEHICLE

LONG VEHICLE

or

LONG VEHICLE

The vertical markings are also required to be fitted to builders' skips placed in the road, commercial vehicles or combinations longer than 13 metres (optional on combinations between 11 and 13 metres)

Hazard warning plates

Certain tank vehicles carrying dangerous goods must display hazard information panels

2YE
1089
Newtown-on-Moors
(0123) 45678

FLAMMABLE LIQUID
3

The above panel will be displayed by vehicles carrying certain dangerous goods in packages

The panel illustrated is for flammable liquid.
Diamond symbols indicating other risks include:

TOXIC 6

Toxic substance

OXIDIZING AGENT 5.1

Oxidizing substance

COMPRESSED GAS 2

Non-flammable compressed gas

RADIOACTIVE 7

Radioactive substance

SPONTANEOUSLY COMBUSTIBLE 4

Spontaneously combustible substance

CORROSIVE 8

Corrosive substance

Projection markers

Side marker

End marker

Both required when load or equipment (eg crane jib) overhangs front or rear by more than two metres

Annexes

1. Choosing and maintaining your bicycle

Make sure that

- you choose the right size of cycle for comfort and safety
- lights and reflectors are kept clean and in good working order
- tyres are in good condition and inflated to the pressure shown on the tyre
- gears are working correctly
- the chain is properly adjusted and oiled
- the saddle and handlebars are adjusted to the correct height.

You **MUST**

- ensure your brakes are efficient
- at night, use lit front and rear lights and have an efficient red rear reflector.

PCUR regs 6 & 10 RVLR no 18

2. Motorcycle licence requirements

If you have a provisional motorcycle licence, you **MUST** satisfactorily complete a Compulsory Basic Training (CBT) course. You can then ride on the public road, with L plates (in Wales either D plates, L plates or both can be used), for up to two years. To obtain your full motorcycle licence you **MUST** pass a motorcycle theory test and then a practical test.

Law RTA 1988 sect 97

If you have a full car licence you may ride motorcycles up to 125cc and 11kW power output, with L plates (and/or D plates in Wales), on public roads, but you **MUST** first satisfactorily complete a CBT course if you have not already done so.

If you have a full moped licence and wish to obtain full motorcycle entitlement you will be required to take a motorcycle theory test if you did not take a separate theory test when you obtained your moped licence. You **MUST** then pass a practical motorcycle test.

Note that if CBT was completed for the full moped licence there is no need to repeat it, but if the moped test was taken before

1/12/90 CBT will need to be completed before riding a motorcycle as a learner.

Law MV(DL)R reg 42(1) & 69(1)

Light motorcycle licence (A1): you take a test on a motorcycle of between 75 and 125cc. If you pass you may ride a motorcycle up to 125cc with power output up to 11kW.

Standard motorcycle licence (A): if your test vehicle is between 120 and 125cc and capable of more than 100 km/h you will be given a standard (A) licence. You will then be restricted to motorcycles of up to 25 kW for two years. After two years you may ride any size machine.

Direct or Accelerated Access enables riders over the age of 21, or those who reach 21 before their two-year restriction ends, to ride larger motorcycles sooner. To obtain a licence to do so they are required to

- have successfully completed a CBT course
- pass a theory test, if they are required to do so
- pass a practical test on a machine with power output of at least 35kW.

To practise, they can ride larger motorcycles, with L plates (and/or D plates in Wales), on public roads, but only when accompanied by an approved instructor on another motorcycle in radio contact.

You **MUST NOT** carry a pillion passenger or pull a trailer until you have passed your test.

Law MV(DL)R reg 16

Moped Licence Requirements

Mopeds are up to 50cc with a maximum speed of 50 km/h.

To ride a moped, learners **MUST**

- be 16 or over
- have a provisional moped licence
- complete CBT training.

You **MUST** first pass the theory test for motorcycles and then the moped practical test to obtain your full moped licence.

If you passed your driving test before 1 February 2001 you are qualified to ride a moped without L plates (and/or D plates in Wales), although it is recommended that you complete CBT before riding on the road. If you passed your driving test after this date you **MUST** complete CBT before riding a moped on the road.

Laws MV(DL)R reg 43

Note. For motorcycle and moped riders wishing to upgrade, the following give exemption from taking the motorcycle theory test
- full A1 motorcycle licence
- full moped licence, if gained after 1/7/96.

Laws MV(DL)R reg 42

3. Motor vehicle documentation and learner driver requirements

Documents

Driving Licence. You **MUST** have a valid signed driving licence for the category of vehicle you are driving. You **MUST** inform the Driver and Vehicle Licensing Agency (DVLA) if you change your name and address.

Law RTA 1988 sect 87

Insurance. You **MUST** have a valid insurance certificate covering you for third party liability. Before driving any vehicle, make sure that it has this cover for your use or that your own insurance gives you adequate cover. You **MUST NOT** drive a vehicle without insurance.

Law RTA 1988 sect 143

MOT. Cars and motorcycles **MUST** normally pass an MOT test three years from the date of the first registration and every year after that. You **MUST NOT** drive a vehicle without an MOT certificate, when it should have one. Driving an unroadworthy vehicle may invalidate your insurance. Exceptionally, you may drive to a pre-arranged test appointment or to a garage for repairs required for the test.

Law RTA 1988 sects 45, 47, 49 & 53

Vehicle Registration Document. Registration documents are issued for all motor vehicles used on the road, describing them (make, model, etc.) and giving details of the registered keeper.

You **MUST** notify the Driver and Vehicle Licensing Agency in Swansea as soon as possible when you buy or sell a vehicle, or if you change your name or address. For registration documents issued after 27 March 1997 the buyer and seller are responsible for completing the registration documents. The seller is responsible for forwarding them to DVLA. The procedures are explained on the back of the registration documents.
Law RV(R&L)R regs 10, 12 & 13

Vehicle Excise Duty. All vehicles used or kept on the roads **MUST** have a valid Vehicle Excise Duty disc (tax disc) displayed at all times. Any vehicle exempt from duty **MUST** display a nil licence.
Law VERA sect 29

Production of documents. You **MUST** be able to produce your driving licence and counterpart, a valid insurance certificate and (if appropriate) a valid MOT certificate, when requested by a police officer. If you cannot do this you may be asked to take them to a police station within seven days.
Law RTA 1988 sects 164 & 165

Learner drivers

Learners driving a car **MUST** hold a valid provisional licence. They **MUST** be supervised by someone at least 21 years old who holds a full EC/EEA licence for that type of car (automatic or manual) and has held one for at least three years.
MV(DL)R reg 16

Vehicles. Any vehicle driven by a learner **MUST** display red L plates. In Wales, either red D plates, red L plates, or both, can be used. Plates **MUST** conform to legal specifications and **MUST** be clearly visible to others from in front of the vehicle and from behind. Plates should be removed or covered when not being driven by a learner (except on driving school vehicles).
Law MV(DL)R reg 16 & sched 4

You **MUST** pass the theory test (if one is required) and then a practical driving test for the category of vehicle you wish to drive before driving unaccompanied.
Law MV(DL)R reg 40

4. The road user and the law

Road traffic law

The following list can be found abbreviated throughout the Code. It is not intended to be a comprehensive guide, but a guide to some of the important points of law. For the precise wording of the law, please refer to the various Acts and Regulations (as amended) indicated in the Code. Abbreviations are listed below.

Most of the provisions apply on all roads throughout Great Britain, although there are some exceptions. The definition of a road in England and Wales is 'any highway and any other road to which the public has access and includes bridges over which a road passes'. In Scotland, there is a similar definition which is extended to include any way over which the public have a right of passage. It is important to note that references to 'road' therefore generally include footpaths, bridle-ways and cycle tracks and many roadways and driveways on private land (including many car parks). In most cases, the law will apply to them and there may be additional rules for particular paths or ways. Some serious driving offences, including drink-driving offences, also apply to all public places, for example public car parks.

Chronically Sick & Disabled Persons Act 1970	CSDPA
Functions of Traffic Wardens Order 1970	FTWO
Highway Act 1835 or 1980 (as indicated)	HA
Horses (Protective Headgear for Young Riders) Regulations 1992	H(PHYR)R
Motor Cycles (Protective Helmets) Regulations 1980	MC(PH)R
Motorways Traffic (England & Wales) Regulations 1982	MT(E&W)R
Motorways Traffic (Scotland) Regulations 1995	MT(S)R
Motor Vehicles (Driving Licences) Regulations 1999	MV(DL)R
Motor Vehicles (Wearing of Seat Belts) Regulations 1993	MV(WSB)R
Motor Vehicles (Wearing of Seat Belts by Children in Front Seats) Regulations 1993	MV(WSBCFS)R
Pedal Cycles (Construction & Use) Regulations 1983	PCUR
Public Passenger Vehicles Act 1981	PPVA
Road Traffic Act 1988 or 1991 (as indicated)	RTA
Road Traffic (New Drivers) Act 1995	RT(ND)A
Road Traffic Regulation Act 1984	RTRA
Road Vehicles (Construction & Use) Regulations 1986	CUR
Road Vehicles Lighting Regulations 1989	RVLR
Road Vehicles (Registration & Licensing) Regulations 1971	RV(R&L)R
Roads (Scotland) Act 1984	R(S)A
Traffic Signs Regulations & General Directions 2002	TSRGD
Vehicle Excise and Registration Act 1994	VERA
Zebra, Pelican and Puffin Pedestrian Crossings Regulations and General Directions 1997	ZPPPCRGD

5. Penalties

Parliament has set the maximum penalties for road traffic offences. The seriousness of the offence is reflected in the maximum penalty. It is for the courts to decide what sentence to impose according to circumstances.

The penalty table, see over, indicates some of the main offences, and the associated penalties. There is a wide range of other more specific offences which, for the sake of simplicity, are not shown here.

The penalty points and disqualification system is described below.

Penalty points and disqualification

The penalty point system is intended to deter drivers from following unsafe driving practices. The court **MUST** order points to be endorsed on the licence according to the fixed number or the range set by Parliament. The accumulation of penalty points acts as a warning to drivers that they risk disqualification if further offences are committed.

A driver who accumulates 12 or more penalty points within a three year period must be disqualified. This will be for a minimum period of six months, or longer if the driver has previously been disqualified.

For every offence which carries penalty points the court has a discretionary power to order the licence holder to be disqualified. This may be for any period the court thinks fit, but will usually be between a week and a few months.

In the case of serious offences, such as dangerous driving and drink-driving, the court **MUST** order disqualification. The minimum period is 12 months, but for repeat offenders or where the alcohol level is high, it may be longer. For example, a second drink-drive offence in the space of 10 years will result in a minimum of three years' disqualification.

Furthermore, in some serious cases, the court **MUST** (in addition to imposing a fixed period of disqualification) order the offender to be disqualified until they pass a driving test. In other cases the court has a discretionary power to order such disqualification. The test may be an ordinary length test or an extended test according to the nature of the offence.

Laws RTRA sects.28,29,34,35 and 36

Penalty table

Offence	Maximum penalties			
	IMPRISONMENT	FINE	DISQUALIFICATION	PENALTY POINTS
*Causing death by dangerous driving	10 years	Unlimited	Obligatory– 2 years minimum	3–11 (if exceptionally not disqualified)
*Dangerous driving	2 years	Unlimited	Obligatory	3–11 (if exceptionally not disqualified)
Causing death by careless driving under the influence of drink or drugs	10 years	Unlimited	Obligatory– 2 years minimum	3–11 (if exceptionally not disqualified)
Careless or inconsiderate driving	-	£2,500	Discretionary	3–9
Driving while unfit through drink or drugs or with excess alcohol; or failing to provide a specimen for analysis	6 months	£5,000	Obligatory	3–11 (if exceptionally not disqualified)
Failing to stop after an accident or failing to report an accident	6 months	£5,000	Discretionary	5–10
Driving when disqualified	6 months (12 months in Scotland)	£5,000	Discretionary	6
Driving after refusal or revocation of licence on medical grounds	6 months	£5,000	Discretionary	3–6
Driving without insurance	-	£5,000	Discretionary	6–8
Driving otherwise than in accordance with a licence	-	£1,000	Discretionary	3–6
Speeding	-	£1,000 (£2,500 for motorway offences)	Discretionary	3–6 or 3 (fixed penalty)
Traffic light offences	-	£1,000	Discretionary	3
No MOT certificate	-	£1,000	-	-
Seat belt offences	-	£500	-	-
Dangerous cycling	-	£2,500	-	-
Careless cycling	-	£1,000	-	-
Cycling on pavement	-	£500	-	-
Failing to identify driver of a vehicle	-	£1,000	Discretionary	3

* Where a court disqualifies a person on conviction for one of these offences, it must order an extended retest. The courts also have discretion to order a retest for any other offence which carries penalty points: an extended retest where disqualification is obligatory, and an ordinary test where disqualification is not obligatory.

New drivers. Special rules apply to drivers within two years of the date of passing their driving test if they passed the test after 1 June 1997 and held nothing but a provisional (learner) licence before passing the test. If the number of penalty points on their licence reaches six or more as a result of offences they commit before the two years are over (including any they committed before they passed the test), their licence will be revoked. They must then reapply for a provisional licence and may drive only as learners until they pass a theory and practical driving test.
Law RT(ND)A

Note. This applies even if they pay by fixed penalty. Drivers who already have a full licence for one type of vehicle are not affected by this when they pass a test to drive another type.

Other consequences of offending

Where an offence is punishable by imprisonment then the vehicle used to commit the offence may be confiscated.

In addition to the penalties a court may decide to impose, the cost of insurance is likely to rise considerably following conviction for a serious driving offence. This is because insurance companies consider such drivers are more likely to have an accident.

Drivers disqualified for drinking and driving twice within 10 years, or once if they are over two and a half times the legal limit, or those who refused to give a specimen, also have to satisfy the Driver and Vehicle Licensing Agency's Medical Branch that they do not have an alcohol problem and are otherwise fit to drive before their licence is returned at the end of their period of disqualification. Persistent misuse of drugs or alcohol may lead to the withdrawal of a driving licence.

6. Vehicle maintenance, safety and security

Vehicle maintenance

Take special care that lights, brakes, steering, exhaust system, seat belts, demisters, wipers and washers are all working. Also

- lights, indicators, reflectors, and number plates **MUST** be kept clean and clear
- windscreens and windows **MUST** be kept clean and free from obstructions to vision

- lights **MUST** be properly adjusted to prevent dazzling other road users. Extra attention needs to be paid to this if the vehicle is heavily loaded
- exhaust emissions **MUST NOT** exceed prescribed levels
- ensure your seat, seat belt, head restraint and mirrors are adjusted correctly before you drive
- items of luggage are securely stowed.

Law: many regulations within CUR cover the above equipment and RVLR regs 23 & 27

Warning displays

Make sure that you understand the meaning of all warning displays on the vehicle instrument panel. Do not ignore warning signs, they could indicate a dangerous fault developing.

- When you turn the ignition key, warning lights will be illuminated but will go out when the engine starts (except the handbrake warning light). If they do not, or if they come on whilst you are driving, stop and investigate the problem, as you could have a serious fault.
- If the charge warning light comes on while you are driving, it may mean that the battery isn't charging. This must also be checked as soon as possible to avoid loss of power to lights and other electrical systems.

Tyres

Tyres **MUST** be correctly inflated and be free from certain cuts and other defects.

Cars, light vans and light trailers MUST have a tread depth of at least 1.6mm across the central three-quarters of the breadth of the tread and around the entire circumference.

Motorcycles, large vehicles and passenger carrying vehicles MUST have a tread depth of at least 1mm across three-quarters of the breadth of the tread and in a continuous band around the entire circumference.

Mopeds should have visible tread.

Laws CUR reg 27

If a tyre bursts while you are driving, try to keep control of your vehicle. Grip the steering wheel firmly and allow the vehicle to roll to a stop at the side of the road.

If you have a flat tyre, stop as soon as it is safe to do so. Only change the tyre if you can do so without putting yourself or others at risk – otherwise call a breakdown service.

Tyre pressures. Check weekly. Do this before your journey, when tyres are cold. Warm or hot tyres may give a misleading reading.

Your brakes and steering will be adversely affected by under-inflated or over-inflated tyres. Excessive or uneven tyre wear may be caused by faults in the braking or suspension systems, or wheels which are out of alignment. Have these faults corrected as soon as possible.

Fluid levels

Check the fluid levels in your vehicle at least weekly. Low brake fluid may result in brake failure and an accident. Make sure you recognise the low fluid warning lights if your vehicle has them fitted.

Before winter

Ensure that the battery is well maintained and that there are appropriate anti-freeze agents in your radiator and windscreen bottle.

Other problems

If your vehicle
- pulls to one side when braking, it is most likely to be a brake fault or incorrectly inflated tyres. Consult a garage or mechanic immediately
- continues to bounce after pushing down on the front or rear, its shock absorbers are worn. Worn shock absorbers can seriously affect the operation of a vehicle and should be replaced
- smells of anything unusual such as burning rubber, petrol or electrical; investigate immediately. Do not risk a fire.

Overheated engines or fire

Most engines are water cooled. If your engine overheats you should wait until it has cooled naturally. Only then remove the coolant filler cap and add water or other coolant.

If your vehicle catches fire, get the occupants out of the vehicle quickly and to a safe place. Do not attempt to extinguish a fire in the engine compartment, as opening the bonnet will make the fire flare. Call the fire brigade.

Petrol stations

Never smoke or use a mobile phone on the forecourt of petrol stations as these are major fire risks and could cause an explosion.

Vehicle security

When you leave your vehicle you should
- remove the ignition key and engage the steering lock
- lock the car, even if you only leave it for a few minutes
- close the windows completely
- never leave children or pets in an unventilated car
- take all contents with you, or lock them in the boot. Remember, for all a thief knows a carrier bag may contain valuables. Never leave vehicle documents in the car.

For extra security fit an anti-theft device such as an alarm or immobiliser. If you are buying a new car it is a good idea to check the level of built-in security features. Consider having your registration number etched on all your car windows. This is a cheap and effective deterrent to professional thieves.

7. First aid on the road

In the event of an accident, you can do a number of things to help, even if you have had no training

1. Deal with danger

Further collisions and fire are the main dangers following an accident. Approach any vehicle involved with care. Switch off all engines and, if possible, warn other traffic. Stop anyone from smoking.

2. Get help

Try to get the assistance of bystanders. Get someone to call the appropriate emergency services as soon as possible. They will need to know the exact location of the accident and the number of vehicles involved.

3. Help those involved

DO NOT move casualties still in vehicles unless further danger is threatened. **DO NOT** remove a motorcyclist's helmet unless it is essential. **DO NOT** give the casualty anything to eat or drink.

DO try to make them comfortable and prevent them from getting cold, but avoid unnecessary movement. **DO** give reassurance confidently to the casualty. They may be shocked but prompt treatment will minimise this.

4. Provide emergency care
Follow the *ABC of First aid*

A is for **Airway** – check for and relieve any obstruction to breathing. Remove any obvious obstruction in the mouth. Breathing may begin and colour improve.

B is for **Breathing** – if breathing does not begin when the airway has been cleared, lift the chin and tilt the head very gently backwards. Pinch the casualty's nostrils and blow into the mouth until the chest rises; withdraw, then repeat regularly once every four seconds until the casualty can breathe unaided.

C is for **Circulation** – prevent blood loss to maintain circulation. If bleeding is present apply firm hand pressure over the wound, preferably using some clean material, without pressing on any foreign body in the wound. Secure a pad with a bandage or length of cloth. Raise the limb to lessen the bleeding, provided it is not broken.

5. Be prepared
Always carry a first aid kit. You could save a life by learning emergency aid and first aid from a qualified organisation, such as the local ambulance services, the St John Ambulance Association and Brigade, St Andrew's Ambulance Association, the British Red Cross or any suitable qualified body.

This Code, between rules 1 and 278, is issued with the Authority of Parliament (laid before both Houses of Parliament June 1998) and appears in the law described as follows:

A failure on the part of a person to observe any provision of **The Highway Code** shall not of itself render that person liable to criminal proceedings of any kind, but any such failure may in any proceedings (whether civil or criminal and including proceedings for an offence under the Traffic Acts, the Public Passenger Vehicles Act 1981 or sections 18 to 23 of the Transport Act 1985) be relied upon by any party to the proceedings as tending to establish or negative any liability which is in question in those proceedings.

Road Traffic Act 1988

Index

References are to rule numbers, except those numbers given in **_bold italic_**, which refer to the annexes

The Highway Code – Index

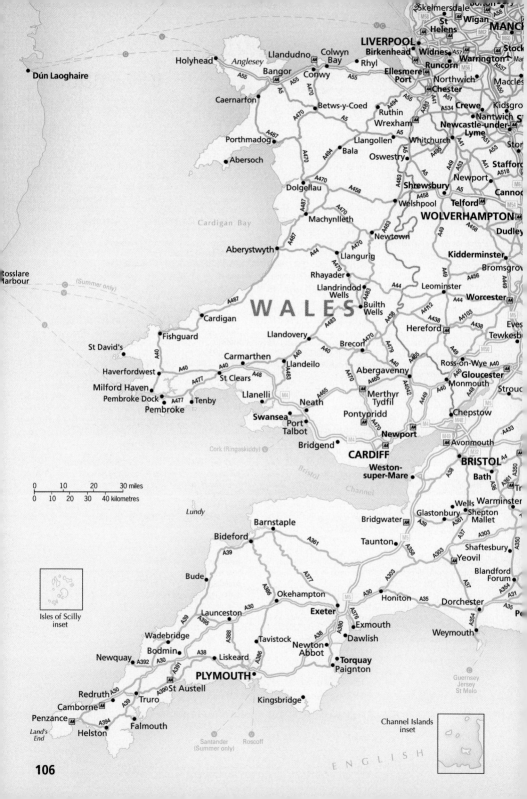

Skelmersdale
St Helens
Wigan
MANCH
LIVERPOOL
Birkenhead
Widnes
Warrington
Stock
Runcorn
Holyhead
Llandudno
Colwyn Bay
Rhyl
Ellesmere Port
Northwich
Maccles
Anglesey
Bangor
Conwy
Chester
Dún Laoghaire
Caernarfon
Betws-y-Coed
Ruthin
Wrexham
Crewe
Kidsgro
Nantwich
Newcastle-under-Lyme
Porthmadog
Llangollen
Whitchurch
Bala
Oswestry
Stor
Abersoch
Shrewsbury
Newport
Stafford
Dolgellau
Welshpool
Telford
Machynlleth
WOLVERHAMPTON
Dudley
Newtown
Aberystwyth
Llangurig
Kidderminster
Bromsgro
Rhayader
Leominster
Worcester
Cardigan Bay
Llandrindod Wells
Builth Wells
Hereford
Eves
Tewkesb
WALES
Cardigan
Llandovery
Brecon
Ross-on-Wye
Gloucester
Rosslare Harbour
Fishguard
Carmarthen
Llandeilo
Abergavenny
Monmouth
Strou
St David's
Haverfordwest
St Clears
Llanelli
Neath
Merthyr Tydfil
Chepstow
Milford Haven
Pembroke Dock
Tenby
Pontypridd
Pembroke
Swansea
Port Talbot
Newport
Avonmouth
Cork (Ringaskiddy)
Bridgend
CARDIFF
BRISTOL
Weston-super-Mare
Bath
Lundy
Wells
Warminster
Barnstaple
Bridgwater
Glastonbury
Shepton Mallet
Bideford
Taunton
Shaftesbury
Yeovil
Bude
Blandford Forum
Okehampton
Honiton
Dorchester
Isles of Scilly inset
Launceston
Exeter
Wadebridge
Tavistock
Newton Abbot
Dawlish
Exmouth
Weymouth
Newquay
Bodmin
Liskeard
Torquay
Paignton
Redruth
St Austell
PLYMOUTH
Guernsey
Jersey
St Malo
Camborne
Truro
Kingsbridge
Penzance
Helston
Falmouth
Channel Islands inset
Land's End
Santander (Summer only)
Roscoff
ENGLISH

AA Driver's
Information

FRANCE

To help you navigate safely
and easily, see the AA's
France and Europe atlases at
www.theAA.com/bookshop

Motorway

Vehicle ferry

Toll motorway

Vehicle ferry -
fast catamaran

Primary route
dual carriageway

Contact your local
AA Service Centre on
0845 603 3111

Primary route
single carriageway

© Automobile Association Developments Limited 2004
Mapping produced by the Cartography Department of the Automobile Association.
© Crown copyright. All rights reserved. Licence number 399221

107

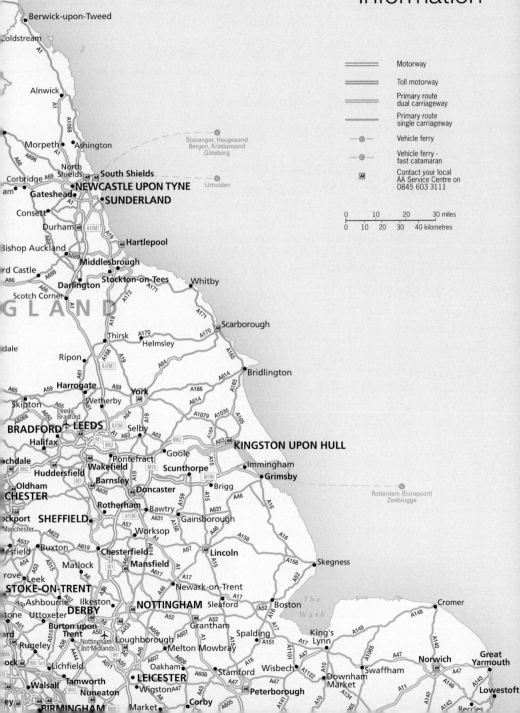

AA Driver's Information

═══════	Motorway
═══════	Toll motorway
═══════	Primary route dual carriageway
───────	Primary route single carriageway
─Ⓥ─	Vehicle ferry
─Ⓒ─	Vehicle ferry - fast catamaran
AA	Contact your local AA Service Centre on 0845 603 3111

```
0        10        20        30 miles
0   10   20   30   40 kilometres
```

Berwick-upon-Tweed
Coldstream
Alnwick
Morpeth • Ashington
North Shields
Corbridge South Shields
am NEWCASTLE UPON TYNE
Gateshead SUNDERLAND
Consett
Durham Hartlepool
Bishop Auckland
rd Castle Middlesbrough
Scotch Corner Stockton-on-Tees Whitby
Darlington
GLAND
Thirsk Scarborough
dale Helmsley
Ripon Bridlington
Harrogate
Skipton York
Wetherby
Leeds
Bradford
BRADFORD LEEDS Selby
Halifax Goole KINGSTON UPON HULL
chdale Pontefract Immingham
Huddersfield Wakefield Scunthorpe Grimsby
Oldham Barnsley Brigg
CHESTER Doncaster
ckport SHEFFIELD Rotherham Rotterdam (Europoort) Zeebrugge
Manchester Bawtry Gainsborough
esfield Buxton Worksop
rove Chesterfield Lincoln
Matlock Mansfield Skegness
STOKE-ON-TRENT Newark-on-Trent Boston
Ashbourne Ilkeston NOTTINGHAM Sleaford The Wash Cromer
rd Uttoxeter DERBY Grantham King's Lynn
tone Burton upon Trent Spalding Norwich Great Yarmouth
Rugeley Nottingham East Midlands Loughborough Melton Mowbray Swaffham
ck Lichfield Oakham Downham Market Lowestoft
ey Walsall Tamworth Wigston Stamford Wisbech Peterborough Beccles
BIRMINGHAM Nuneaton LEICESTER Corby Market
Stavanger, Haugesund Bergen, Kristiansand Göteborg
IJmuiden

Port of Ness

Western Isles

Stornoway

Outer Hebrides

The Minch

Isle of Lewis

Tarbert

Ullapool

Harris

Gairloch

A835

North Uist

Lochmaddy

Uig

A87

Benbecula

South Uist

Isle of Skye

Portree

Kyle of Lochalsh

Inve

Lochboisdale

A87

A87

A887

Barra

A87

Invergarry

A86

Rùm

Mallaig

Eigg

A830

A82

Inner Hebrides

Fort William

Coll

A82

Tiree

Isle of Mull

A828

SCO

Oban

A85

A85

A816

Crianlar

Colonsay

A83

A82

Jura

Helensburgh

Dunoon

Greenock

Cl

A78

Glasgow

Paisley

Tarbert

Largs

Islay

A737

Port Ellen

A83

Irvine

A71

A77

Troon

A76

Kilmar

Arran

Ayr

A77

A70

Campbeltown

Firth of Clyde

Girvan

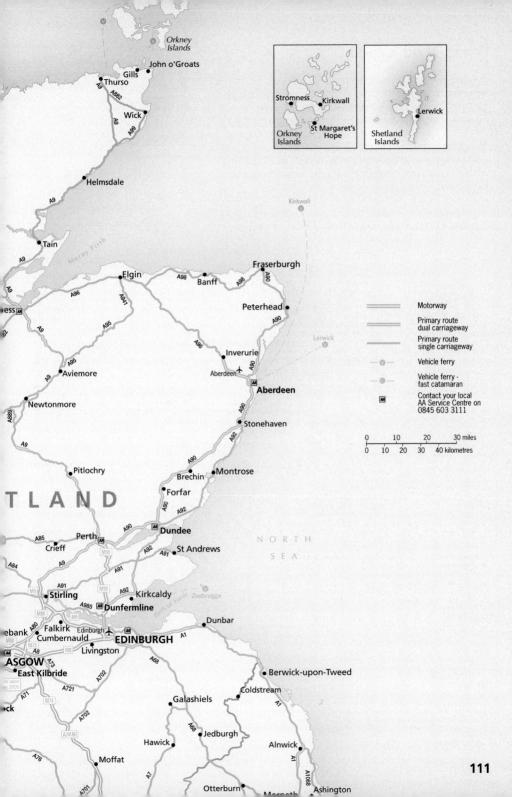

Orkney
Islands

Gills
John o'Groats
Thurso
Wick

Stromness
Kirkwall
Orkney
Islands
St Margaret's
Hope

Lerwick
Shetland
Islands

Helmsdale

Kirkwall

Tain

Moray Firth

Elgin
Banff
Fraserburgh

Peterhead

INESS

Aviemore

Inverurie
Aberdeen

Newtonmore

Aberdeen

Stonehaven

Pitlochry

Brechin
Montrose
Forfar

TLAND

Perth
Crieff
Dundee

NORTH
SEA

St Andrews

Stirling
Kirkcaldy
Dunfermline

Zeebrugge

ebank
Falkirk
Edinburgh
Cumbernauld
Dunbar

ASGOW
East Kilbride
Livingston
EDINBURGH

Berwick-upon-Tweed
Coldstream

Galashiels

Jedburgh
Hawick
Alnwick

Moffat

Otterburn
Ashington

Motorway

Primary route
dual carriageway

Primary route
single carriageway

Vehicle ferry

Vehicle ferry -
fast catamaran

Contact your local
AA Service Centre on
0845 603 3111

0   10   20   30 miles
0   10   20   30   40 kilometres

111

Mileage chart

The mileage chart shows distances in miles between two towns along AA-recommended routes. Using motorways and other main roads this is normally the fastest route, though not necessarily the shortest.

The journey times, shown in hours and minutes, are average off-peak driving times along AA-recommended routes. These times should be used as a guide only and do not allow for unforeseen traffic delays, rest breaks or fuel stops.

For example, the 378 miles (608 km) journey between Glasgow and Norwich should take approximately 7 hours 28 minutes.

journey times

(Diagonal mileage/journey-time chart listing distances in miles and journey times between towns including: Aberdeen, Aberystwyth, Barnstaple, Birmingham, Brighton, Bristol, Cambridge, Cardiff, Carlisle, Carmarthen, Dorchester, Dover, Edinburgh, Exeter, Fort William, Glasgow, Gloucester, Guildford, Hereford, Holyhead, Hull, Inverness, Kendal, Leeds, Lincoln, Liverpool, Maidstone, Manchester, Middlesbrough, Newcastle, Northampton, Norwich, Nottingham, Oxford, Penzance, Perth, Peterborough, Plymouth, Portsmouth, Preston, Salisbury, Sheffield, Shrewsbury, Southampton, Stoke-on-Trent, Stranraer, Taunton, Wick, York, LONDON.)

distances in miles (one mile equals 1.6093 km)